FOREWORD

This book is a collection of the written papers presented and discussed at the ASME Internal Combustion Engine Division's 1991 Fall Technical Conference held September 30- October 3, 1991 at the Holiday Inn in Muskegon, Michigan. The ICED Fall Technical Conferences were initiated in 1979 to complement the ETCE-ICED Symposium and to provide a forum to share technological advancements in internal combustion engines. These Fall Technical Conferences have a timely theme around which the technical papers and presentations center. Past conferences have dealt with engine efficiency, alternate fuels, lubricants, heavy petroleum fuels, electronics, cogeneration, emissions technology, computers in design, operation and controls, and new technology in large bore engines. This year's theme is "Fuels, Controls, and Aftertreatment for Low Emission Engines".

There is renewed interest in the area of engine emissions due to 1990 Clean Air Act Amendments and also because many urban areas throughout the world are struggling to control emissions from various man made sources. The engineering and scientific community is therefore investigating the use of alternate fuels for engines. The papers in this book present some of the results from these efforts. If the engine out emissions are not acceptable, then aftertreatment becomes the alternate choice; papers covering this subject are included in this book. This book also contains papers on the electronic controls technology required to obtain the optimum performance from the engine. All the above areas have been appropriately tied together in the Soichiro Honda Lecture "Energy, Efficiency, and the Environment - Three Big Es for Transportation".

All papers in this booklet have been peer reviewed by at least two persons and conform to standard ASME publication practices.

I would like to extend thanks to the authors for sharing their research and insight with the engineering and scientific community and the many persons who assisted in the review process. I would also like to thank Paul Blumberg, Jerry Caton, Jim Garrett, Ram Sharma, Terry Ullman and Teoman Uzkan for their efforts in organizing the conference.

Madan R. Goyal

CONTENTS

KARL J. SPRINGER

1991 SOICHIRO HONDA LECTURER

Karl J. Springer will present the fifth Soichiro Honda Lecture on September 30, 1991, at the ASME ICE Fall Technical Conference in Muskegon, Michigan. He has selected, "Energy, Efficiency, and the Environment - Three Big Es of Transportation" as the subject for the 1991 Soichiro Honda Lecture.

Mr. Springer is Vice President of the Automotive Products and Emissions Research Division of Southwest Research Institute. Karl did pioneering research in developing procedures and equipment for the measurement of diesel smoke and odor. He has made significant contributions in the area of diesel engine particulate collection and measurement including a major role in the development of the EPA Transient Test Procedure. He has also worked extensively in the areas of fuels and lubricants research.

Karl has published widely on the subjects of emissions, fuels, lubricants, and the environment, having been author or co-author of over 37 technical papers. Three of these were selected for Outstanding Paper Awards by this Division. Karl is an enthusiastic participant in the Internal Combustion Engine Division, having recently served as its chairman. He is a Fellow of both ASME and SAE, and has received wide recognition.

The Soichiro Honda Lecture has been established as a National Lecture by ASME to recognize achievement and significant contribution in the field of personal transportation. It is a highlight of the ICE Division Fall Technical Conference. Past recipients are Helmut List, President of AVL, Dr. Phillip Myers, Emeritus Professor, University of Wisconsin, Dr. Horst Hardenberg, Director of Advanced Truck Engine R&D, Daimler-Benz, and Professor John B. Heywood, Director of the Sloan Automotive Laboratory and Professor of Mechanical Engineering at Massachusetts Institute of Technology.

ICE-Vol. 15, Fuels, Controls, and Aftertreatment
For Low Emissions Engines
ASME 1991

1991 SOICHIRO HONDA LECTURE

Energy, Efficiency, and the Environment: Three Big Es of Transportation

Karl J. Springer
Southwest Research Institute
San Antonio, Texas

ABSTRACT

The three big Es of transportation are Energy, Efficiency, and the Environment. As the clouds of global climate change and the desire to rely less on Mid East crude builds, how do the manufacturers and refiners reconcile the needs of consumer acceptance and governmental regulation? How can policies and practices be united so that everyone involved works to the common goal of personal mobility? This lecture traces recent events that have resulted in paradox on top of paradox. As industry continues to react to the latest round of air pollution regulations, where are we headed as far as new CAFE limits and the potential for additional longer term controls related to the greenhouse effect? These are issues that will affect those in the equipment and oil industry, as well as the consumer, in the days ahead.

INTRODUCTION

This is a comment on current and future road transportation, as influenced by sometimes contradictory and paradoxical policies of government and responses by industry. In no other major segment of our society have government regulations roamed at will than in on-highway transportation, regulating everything from the fuels to the tailpipe emissions with safety and fuel economy thrown in for good measure. For the past 20 years, the thrust of these regulations has been principally at the car makers. Because car making and fuel refining are very competitive businesses, the industry had to have common targets to justify the enormous redirection of effort and additional expenditures to reduce energy consumption and emissions. The threat of "you cannot sell cars in the U.S. after a given date unless they met certain emission, safety or economy requirements" got the attention of upper management. Perhaps this approach was necessary to get action on items generally considered as additional costs to make and sell a product, but were not perceived as a reason for a customer to decide to buy that make of car. Therein was the great argument which deeply divided the industry from the government. The car and equipment making and fuel refining industries resented the government and its regulators intruding into their business. They resented the agency employees telling them how to make cars, trucks, and buses, and how to refine crude oil. It made little difference that the Environmental Protection Agency (EPA) or Department of Transportation (DOT) were, in fact, carrying out the wishes of Congress. In many cases, the Federal law was re-interpreted through Federal court decisions as a result of law suits by one or more environmental or consumer groups.

In retrospect, the industry has been able to comply with these numerous regulations albeit at enormous cost in resources of people, equipment, and facility; a cost eventually passed to the consumer. Only in a few instances were time delays necessary. Thus, the lawmakers and regulators can take some measure of satisfaction in saying, "See, I told you you could do it" and "my technology-forcing regulations were achievable after all." There is little doubt that we now have more environmentally friendly, safe, and efficient cars at an earlier date than we would have had if the industry had been left to its own timetable. Perhaps this is the one saving feature, the one redeeming value in the system of confrontation, regulatory requirements and reactive response that characterized each of the areas of environment, fuel, and safety improvements. It would be sufficient to end this lecture at this point if the saga of regulation of the automotive industry was at a stand still, as it was during the eighties. Everyone could take credit for winning, in their own way, since everyone won; manufacturer, refiner, and consumer. The products are definitely better, and the consumer did not have to pay as much as some forecast he would, even though he is paying more.

There are new U.S. and state laws, however, that portend enormous change to transportation that affect makers, refiners, and consumers in the future. It is these new requirements and the changes, conflicts, contradictions, and paradoxes that need debate and discussion. Fresh from that extra good feeling of accomplishment by the Clean Air Act amendments of 1970 and 1977 and the Energy Policy and Conservation Act of 1975 establishing Corporate Average Fuel Economy (CAFE), the lawmakers have tasked the industry with even more technology-forcing requirements for the decade of the nineties and beyond. It is these forces, clearly the drivers of technology improvement for the foreseeable future, that will shape the types of products to be used by consumers in the U.S. and to some extent in other parts of the world. So far, safety regulations for the future have concentrated on extending the passenger car rules to other classes of motor vehicles. Safety remains important, however, as the car is made even smaller and lighter, or must use a new fuel or an existing fuel under different circumstances. The remainder of this lecture will focus on energy, efficiency, and environmental issues, the three big "Es" or "biggies," as they are expected to affect on-highway cars, trucks, and buses.

ENERGY

The automotive car, truck, and bus engine has been married to crude oil derived gasoline and diesel fuel almost since inception. Development and use of the internal combustion engine powered vehicle has been linked directly with a relatively inexpensive, abundant and available refined product. This marriage of fuel and engine has had its rocky times, as in 1967 when Exxon[1]* showed the auto industry that activated carbon could serve to collect vapors from the tank and engine. This resulted in a Federal Regulation[2] to control evaporative emissions during diurnal heating as well as after engine shut-off. Another instance was when, at the request of the auto makers, the EPA required unleaded gasoline for catalyst-equipped cars beginning July 1974. Yet another instance was in 1985 when General Motors urged fuel refiners to make cleaner gasoline that would not make troublesome port fuel injector deposits[3]. Sometimes, the relationship seemed troubled, marked by lengthy silences, as if the partners had quit talking to each other. The marriage appeared headed for the divorce court when an interloper, alcohol fuel, entered the scene with strong Federal encouragement in June 1989. The car companies had developed flexible fuel vehicles (FFVs) able to run on any combination of methanol and gasoline, up to 85 percent by volume methanol, called M-85.

Auto-Oil Fuel Effects Project

In October 1989, the 14 million dollar, Phase I Auto-Oil Program of 1990-1991 was announced[4]. Perhaps the turning point was the June 1989 Bush Administration proposal[5] to mandate millions of FFVs in nine non-attainment areas on a time-phased schedule. Maybe the turning point was the September 1989 introduction of a reformulated fuel by ARCO called EC-1 to replace leaded gasoline in Southern California[6]. Perhaps there was the growing realization of a need to see what might be possible with gasoline composition in conjunction with the modern computer-controlled passenger car[7]. In November 1990, Phase II of the auto-oil program was authorized. It is a 25 million dollar project expected to be completed in mid-1994.

There was little doubt by anyone that the automobile had made truly remarkable, revolutionary, gains in all the regulated areas of emissions, economy, and safety, while still burning the same basic unleaded gasoline for the past 15 years. Was it time for the possibilities of improved gasoline to be revisited? It had been over 20 years since the last in-depth research on fuel effects on emissions and air quality had been performed. The 1970 Clean Air Act Amendments directed major reductions in tailpipe emissions with goals of 90 percent HC and CO and 70 percent NO_x. The oil industry was left out of such technology forcing limits for good reason. Changes in fuels, while producing some reductions in some pollutants by perhaps 10-15 percent were trivial compared to the 70 and 90 percent requirements placed on the manufacturer. As a consequence, little was asked from the refiner other than to take out the lead (starting in July 1974) and add deposit control additives (1985) and control vapor pressure (1989)[8].

*[1] References listed at the end of this paper.

Energy Consumption

What does all this have to do with energy other than gasoline and diesel fuel are made from crude oil? As shown in Figure 1[9], approximately 11 million barrels of crude oil are consumed by transportation in the U.S. each day. About 45 percent is imported, affecting more or less the U.S. balance of trade. These statistics do not mean much until one realizes that it was transportation, principally the popularity of the automobile, that increased our need for crude oil. The U.S., it seems, will do almost anything to get oil. How different the U.S. and the world might be had the personal freedom of mobility provided by the privately owned car been restricted by less availability of crude oil. The availability of crude oil to the U.S. market was somewhat dependent on the advent of cheap transport in huge supertankers, the technical innovation which permitted oil from relatively inaccessible sources a way to reach the U.S. market. Post World War II government policies that supported urban sprawl, endless highway construction, the suburbs, and low density housing away from the city[10] also take much credit for the car becoming a necessity to life, society, and civilization as we know it today.

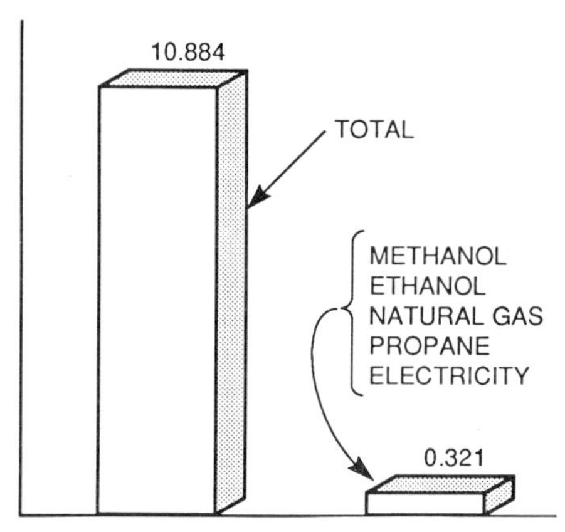

Figure 1 Transportation Energy, Estimated 1989 U.S. Daily Average (Millions of Barrels Per Day Oil Equivalent)[9]

The major driving force for the mixed fuel-energy strategy of today is concern over environmental effects, termed micro and macro. It happens that nearly every city of over 1 million people has some type of air pollution problem dependent more or less on transportation along with the other activities of the city. This is termed the micro effect. Relative to the seven mile thickness of the troposphere, the layer of atmosphere that surrounds the planet Earth, the city-related pollution is regarded by some as isolated. Depending on the gas or vapor, however, all have some life in the atmosphere regardless of whether generated in the city or in the country. This is the macro effect.

Carbon Based Fuels - Global Warming

Consumption of energy, principally from fossil fuel, means a carbon-based economy; one in which the standard of living is proportional to the amount of carbon consumed, whether from wood, charcoal, coal, oil, or natural gas. The increasing population of the Earth, the desire for a better life style, including all the material goods and services, plus the freedom of personal mobility afforded by the car, means more carbon consumption, at an ever-increasing rate. Figure 2[11] shows that more and more CO_2 is released as the carbon in the fuel is burned whether in a boiler or in an internal combustion engine.

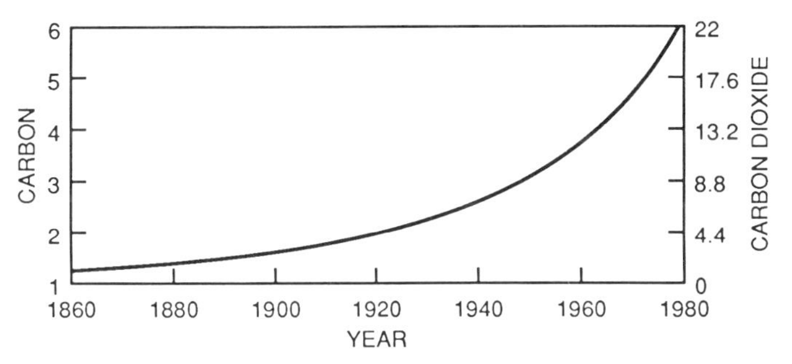

Figure 2 Worldwide Carbon and Carbon Dioxide from Fossil Fuel Combustion (Billions of Tons Per Year)[11]

An ever-increasing CO_2 concentration gives rise to concern over global warming and the potential for global climate catastrophe. Carbon dioxide, along with methane, chloroflourocarbons (CFCs), nitrous oxide, tropospheric ozone and water vapor, absorbs the reflected infrared energy from sunlight that strikes the Earth. Like panes of glass in a greenhouse, the absorption of this energy is responsible for a livable earth environment of about 15°C (59°F). Too much greenhouse gas emissions portend disaster according to a growing number of scientists. Since CO_2 is believed to be half of the greenhouse gas effect, the ever-increasing consumption of carbon-containing fuels is suspect and is being scrutinized by world opinion. In the U.S., transportation accounts for 19 percent of the CO_2, while in the world, transportation accounts for 4.2 percent, of which cars are responsible for 2.8 percent of the CO_2 emissions.

More than 130 nations took part in the February 4-14, 1991 Chantilly, Virginia International Negotiating Committee on a Framework Convention on Climate Change[12]. It is expected that in 1992, the U.S. may enter into some type of international agreement on limiting CO_2 emissions. When this happens, it will likely include transportation and especially the passenger car, a favorite topic for CO_2 reduction. There are many strategies under consideration which involve, in the ultimate, a move away from carbon-containing gasoline, diesel, or alcohol fuels to electricity and hydrogen made from non-CO_2 producing sources of energy such as the sun, nuclear, or wind. The staggering amount of energy used each day in the U.S. for transportation makes any alternative look like the proverbial "drop in the bucket." Thus, such transportation strategies will likely include major efforts to conserve, in terms of improved efficiency, reduction in vehicle miles traveled as in a change in life style and perhaps limitations in auto registrations or fuel rationing. Much has been published on this subject and references 13 through 16 are helpful for further study.

EFFICIENCY

There are some common misconceptions about fuels and what influence environmental rules have had and continue to have over engine efficiency as well as the ubiquitous corporate average fuel consumption (CAFE) regulations.

Spark-Ignited Engines

The spark-ignited (Otto cycle) engine thermal efficiency is dependent on compression ratio among other factors. Figure 3, from reference 17, shows the historical trend of compression ratio in the U.S. Spark ignition engine compression ratio is determined by the mechanical octane number of the combustion chamber design such as high swirl, fast-burn, systems and is limited by the fuel quality or chemical octane number. In the design of engines that have greater efficiency, gasoline with higher octane is essential. An engine's tendency to knock audibly under low speed, relatively mild, conditions is measured by ASTM D 2699 for research octane number (RON) and under high speed, high severity, conditions by ASTM D 2700 for motor octane number (MON).

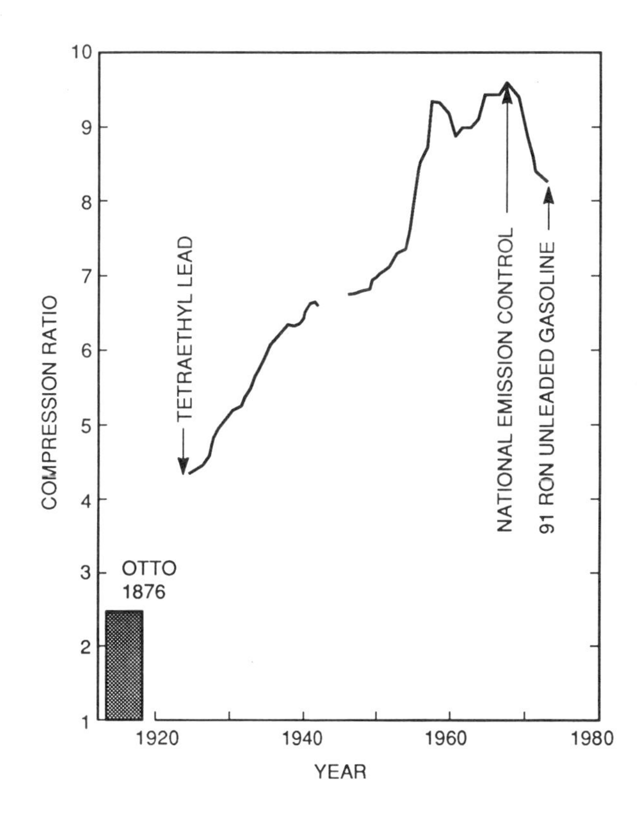

Figure 3 Average U.S. Compression Ratio, Historical Trend[17]

Sometimes cars will have low-speed knock during light acceleration or knock audibly during hard acceleration as in entering a freeway or during hill climb. Usually one cylinder is responsible because of excessive combustion chamber deposits, over advancement of the spark timing, or some change in EGR or air-fuel mixture. Such knock is of limited duration and is rarely harmful to the engine. The use of higher octane, premium grade, gasoline usually eliminates the annoying knocking sound, but does little to improve the engine thermal efficiency or vehicle fuel economy, contrary to popular belief. EPA has determined that only 5 percent of the cars actually need premium or mid-grade unleaded, although those grades are as much as one-third of the gasoline sold.

In 1921, tetraethyl lead (TEL) was shown to decompose during the combustion cycle to form a cloud of catalytically active lead oxide particles. These particles interrupt the chain-branching reactions which lead to too-rapid combustion known as knock. The maximum lead content of gasoline was initially established by the U.S. Surgeon General at 4.0 milliliters of lead compound per gallon (4.23 grams of lead per gallon) based on health considerations in handling the fuel. The economical use of lead resulted in a practical maximum of about 3 grams per gallon. The effectiveness of TEL and TML lead compounds are dependent on fuel composition. TEL, and more recently TML (tetramethyl lead), with appropriate bromine and chlorine based scavengers, have been widely used in gasoline in the U.S. Considerable advances were also made in refining processes such as thermal cracking, catalytic cracking, and reforming, so that by about 1955, 95 RON gasoline was available.

Compression Ignited Engines

This type of engine, which normally operates on diesel fuel, is the most efficient of any heat engine yet devised. Diesel fuel has an ignition quality, called cetane number, which is a measure of the readiness of the fuel to ignite spontaneously. It is opposite in effect to that of the octane quality of gasoline. The higher the cetane number, the shorter the delay between injection and ignition of the fuel under the pressure and temperature in the combustion chamber of the engine. Fuel cetane number for type 2-D fuel, popularly used in on-road trucks and buses, is specified by ASTM D 975-81 to be 40 minimum, as measured by ASTM D 613. Sometimes low cetane levels result in increased combustion noise, emissions, and poor starting. The effect of cetane on efficiency is more dependent on the fuel composition and heating value with an average U.S. cetane of 45, common for on-highway type 2-D fuel. Merely increasing cetane number does not necessarily allow a higher compression ratio or result in lower fuel consumption. Diesel engines use a compression ratio high enough for starting, yet optimum for best engine efficiency and durability. One compression ignition engine that runs on methanol features a compression ratio as high as 23:1, plus ignition aids such as glow plugs because of the much higher auto-ignition temperature of methanol relative to diesel fuel[18].

Air Pollution and Unleaded Gasoline

During the fifties and sixties, air pollution over some cities in the U.S., notably Los Angeles, got progressively worse as photochemically reactive hydrocarbons and oxides of nitrogen combined in sunlight to yield ozone and cause eye irritation. The best strategy for meeting the stringent requirements of the 1970 Clean Air Act Amendments was to reduce engine-out emissions as much as possible and then add an oxidation catalyst to further reduce CO and HC. Retarded ignition timing and exhaust gas recirculation were popularly used ways to reduce NO_x.

In order to use the oxidation catalyst, the car makers needed a lead-free fuel. It was known that lead combustion products glaze the porous surface of the exhaust catalyst, a poisoning effect that rendered the device less efficient. The U.S. car makers, led by GM President Edward Cole, redesigned engines for 1975 with lower compression ratio. In this way, the engines could operate on regular grade gasoline without lead. On January 10, 1973, the EPA issued regulations requiring unleaded gasoline be sold starting July 1, 1974 for catalyst equipped cars. The unleaded gasoline was to have 91 RON, later defined as octane index (R+M)/2 of 87.

The car makers did not ask the oil refiners to change composition or reformulate, merely take out lead. This was not easily accomplished, however, since lead was a blending agent that was not used to the same amount in all gasolines; nor was the phase-in of unleaded cars all that the refiner had to deal with. The EPA soon issued rules to phase-out lead content in the remaining gasoline beginning January 1, 1975, which did result in changes in refinery processing and fuel composition, now known as reformulation. Many times, these changes adversely affected refinery yield, requiring more energy consumption to produce more and more unleaded or lower lead gasoline; another loss in the overall energy efficiency. Some of the engine efficiency loss was regained through continued combustion research, improved mixing and combustion process control, an improvement in mechanical octane number. In some cases refinery yields likewise were improved. However, the advent of lower compression engines and the refining of unleaded gasoline are still considered to have increased overall energy consumption, a paradox of some significance.

OPEC, the Energy Crisis and CAFE

In 1973, OPEC limited production to increase the price of crude oil. OPEC said it was trying to do us a favor with higher prices to reduce consumption and become more self reliant. In the energy crisis which resulted, the CAFE regulations required makers of cars and light trucks to produce more fuel efficient vehicles ultimately reaching 27.5 miles per gallon (MPG) average in 1985. Figure 4[19] tracks the sales weighted CAFE, in miles per gallon, since 1974 for both domestic and imported cars. Interestingly, the engine design improvements played a minor role although engine power to vehicle weight were of major importance. The tried-and-true methods to achieve better economy through lower vehicle weight, smaller displacement engines; i.e., 4 cylinder in place of 6 and a 6 in place of a V8, lower engine speeds, by use of overdrive transmissions, and lower wind and rolling resistance all helped. At no time did the fuel itself become a subject of reformulation or change. Compression ratios remained at a level compatible with (R+M)/2 of 87 (91 RON). There was little thought given to raising fuel octane and compression ratio. With the crisis conditions of shortages, gas lines, and higher prices, people were glad there was fuel enough to burn. Alternatives such as diesel fuel made a brief appearance in U.S. made passenger cars as a means to preserve the large car in some vehicle line-ups. Diesel cars have superior fuel economy; however, few diesel powered cars are still sold in the U.S. because of the stringent exhaust particulate standards.

Chart tracks sales-weighted corporate average fuel economy for domestic and imported car fleets. Figures in miles per gallon.

	'74	'75	'76	'77	'78	'79	'80	'81	'82	'83	'84	'85	'86	'87	'88	'89
IMPORTS	22.2	23.2	25.4	27.7	27.3	26.1	29.6	31.5	31.1	32.2	31.8	31.5	31.7	31.1	31.2	30.6
DOMESTIC	13.2	14.8	16.6	17.2	18.7	19.3	22.6	24.2	25.0	24.6	25.6	26.3	26.6	27.0	27.3	27.1

Source: U.S. Department of Transportation

Figure 4 CAFE Since 1974[19]

At the same time the safety standards and CAFE standards were taxing the car makers, another, more significant, reduction in emissions was promulgated. As a result of the Clean Air Act Amendments of 1977, the car maker needed to rollback HC, CO, and NO_x emissions even more. Effective with model year 1981, HC/CO/NO_x limits were 0.41/3.4/1.0 grams per mile. To do so required a major technical break-through, the oxygen sensor. This sensor permitted use of a three-way catalyst, one which simultaneously could promote oxidation of HC and CO to CO_2 and water vapor and reduction of NO to N_2, water vapor, and CO_2. To make this happen, the oxygen sensor acted as a fast-acting switch at the stoichiometric air:fuel ratio, about 14.7:1 for gasoline, to provide feedback to the engine. It happens that the use of platinum (for oxidation) and rhodium (for reduction) in the same catalyst system would reduce all three regulated emissions if the air:fuel ratio was controlled to 14.7:$\pm$0.1. Unleaded gasoline remained a requirement for long life of both the porous zirconium dioxide oxygen sensor and the catalytic device. A variety of system controls for fuel and air resulted. With individual port fuel injectors directing a fuel spray on the underside of the intake valve and utilizing the error signal from the oxygen sensor for feedback, the possibilities of computer control of the engine continue to be explored to achieve long-term driveability and customer acceptance. Such systems, considered revolutionary just a few years ago, are considered commonplace and even taken for granted today.

Now possible are sensors that detect knock before it can be heard and automatically determine which cylinder is knocking. Using a smart ignition system, the timing of just the knocking cylinder is retarded for the usually brief knocking duration and then readjusted back to standard programmed timing. This engine of today has the ability to be octane insensitive; i.e., run on a fairly wide range of fuels. Of course, the lower the octane, the later the ignition in more and more cylinders and although the engine will still run, the performance and efficiency will be poor. Too bad these same smart engines cannot advance the timing to take advantage of the higher octane in some fuels. Rarely is this permitted because the increase in engine-out NO could overload the 3-way catalyst. The engine featuring manifold fuel injection, oxygen sensor, and 3-way catalyst with adaptive feedback computer control and knock sensor has a trivial effect on engine efficiency, even though there are momentary retarded timings where the occasional cylinder knock occurs.

To some engineers, the 3-way catalyst represents a major obstacle to improved efficiency, an improvement of as much as 10 percent promised from very lean-burning engines. Engines that operate with air:fuel ratios of greater than 22:1 qualify as lean burn and are possible with today's technology. The fact that 3-way stoichiometric burn engines are already available and in common use for many years in the U.S. has been reason enough for the European community to adopt the same stringent standards for its member countries. These standards are sufficiently difficult to rule out, at least for the time being, serious consideration of the lean burn engine. So far, lean burn engines cannot achieve the NO_x standard and being lean, cannot make use of 3-way catalytic aftertreatment because of the excess oxygen in the exhaust. Such regulations are seen by the European and Japanese to deter and perhaps block further development of lean burn engines, another obvious efficiency-environmental paradox.

Optimum Octane - Efficiency & Energy Use

It requires more energy to make gasoline to the same octane number without benefit of TEL or TML. The current unleaded 91 RON level is not necessarily optimum for overall best energy use in the refinery and in the engine. For example, a 95 RON fuel is sufficient to allow the compression ratio to be increased a full number from say 8.5 to 9.5:1. However, it is pointless to make higher compression engines achieve better combustion efficiency if the gains are more than offset by efficiency losses in the refinery. A 1983 study on the rational utilization of fuels in private transport[20] indicated 94.5 RON is optimum. Similar studies in the U.S. need to take advantage of the latest technical improvements in refinery processes and engine possibilities. The optimum fuel octane value could help the auto-oil program to define a recipe for future gasoline and, of course, be central to any energy plan or policy. A "rule of thumb" is an increase in one octane number will permit a one percent fuel economy improvement.

Future Efficiency Improvements - The major improvements in engine efficiency, expected in the future, will lay with reduced friction losses, reduced pumping losses, preservation of heat through insulating films and coatings coupled with improved expansion processes[21]. Heat conservation within the engine does little good if it is not converted to useful work in the expansion process or collected in some type of regenerative turbo-compound expander, where exhaust heat is converted into torque. Of all these potential areas, the remaining major improvements in engine efficiency are thought to be related to engine friction. Significant efforts are in progress and more, much more, is expected in terms of design, materials, and lubricants. Even so, all these changes will be minor and almost trivial when compared to further improvements in engine and vehicle weight, vehicle power to weight, size, and shape, as the car maker struggles to keep gasoline cars in the mix of vehicles for the expected future stringent CAFE regulations.

Safety concerns already are voiced as vehicle size is further reduced. The small car injury statistics and the concern over vehicle side and head-on collisions may place some lower limit of vehicle size and weight for safety reasons. If all cars were underpowered and the same weight, then the problem would not be as severe. This is obviously not going to be the case in the near future any more than in the past. The car designer will be forced to the limit of technology in the use of materials and structures for occupant protection as vehicles are further reduced in weight[22].

Another major area for continued improvement is in drive train torque management. There is a limit to how slow the engine can run with acceptable torque. However, the more speeds in the automatic transmission, the greater the ability of the computer to match the transmission to the engine for the drive torque required. Computer management of engine and transmission will lead to further improvements in the quality of drive, the ultimate test of customer satisfaction, and this is affected by the fluid as well as the friction materials. It is believed that the computer controlled automatic transmission passenger car can achieve superior fuel economy to the manual transmission except when driven by an expert test driver. The computer control will consistently decide on the best drive gear for fuel economy and driveability and thus be important in achieving future CAFE limits.

Regardless of all these expected improvements in efficiency, the gasoline-powered car will find itself in a mix of vehicles powered by other types of fuels and engines. The mix could include FFVs or vehicles with engines designed to burn methanol only or natural gas. Depending on future CAFE regulations and provisions to give credits to alternative fuels, electric, gasoline-electric hybrids, and even hydrogen-fueled vehicles could appear. One suggestion that encourages alternatives is to compute fuel economy on the basis of the amount of crude oil derived gasoline consumed. For a FFV operating on M-85, 15 percent gasoline, 85 percent methanol, getting 20 mpg (gasoline energy equivalent), its fuel economy could be 133 mpg if based only on the gasoline fraction[23]. Such calculations presume that the FFV will always operate on M-85, an uncertain possibility.

Future of CAFE - Major increases in CAFE are expected to make major changes in vehicles. This is thought to be the single most important issue facing the auto makers. As mentioned earlier, there are still some technical improvements possible. However, these will not be, in themselves, sufficient to meet improvements proposed by Senator Bryan of 20 percent (to 34 mpg) and 40 percent (to 40 mpg) in 1996 and 2001[24], or Senate Bill 279, a bipartisan bill to increase CAFE to 40 mpg by 2001. CAFE should be considered as one of several aspects of conservation. To some, it is the only "safe" or convenient policy to follow regarding reduced reliance on foreign crude oil, global warming, trade imbalance, and air pollution. To others, it is a way to encourage alternatives through creative fuel economy incentives.

Those against CAFE regulation argue that the best policy would be to sharply increase the price of gasoline through taxation, as do many European countries. Others believe a reduction in vehicle miles traveled by increasing the number of passengers per car from the current average of about 1.5 is the answer. This strikes at a life style based on freedom of mobility and the cost and convenience of personal vehicle operation. Since no one wishes to interfere in these, the major burden of fuel economy improvements continue to be the responsibility of the car maker.

Heavy Duty Engine and Vehicle Efficiency

EPA classifies as heavy duty those engines used in trucks and buses having a gross vehicle weight (GVW) of over 8500 lbs. or a frontal area larger than 45 ft^2. Such engines and vehicles have no fuel economy or CAFE standards. The value of HD engines is in their ability to produce flywheel power and torque, and not carry on the average of 1.5 persons, as with cars and many pick-up trucks. Engines usually are specified on the basis of specific fuel consumption (lbs. of fuel per bhp hour) at rated speed and load, useful in comparing engines such as gasoline and diesel. The energy crisis of the mid-seventies brought out many new engine and vehicle features such as engine speed derating, greater use of turbochargers, wind deflectors for less wind resistance, and lower rolling resistance tires. To what extent future regulations will impact on truck and bus fuel economy over typical driving cycles, that are somewhat analogous to the passenger car city and highway cycles, is unknown. The EPA transient engine certification cycle results in fuel consumption at the same time the emissions of HC, CO, NO_x and particulate (diesel only) are obtained. However, the analogy with the CAFE procedure ends there since engine fuel consumption is only one part of HD vehicle fuel consumption, a value not readily obtained as is the case for light duty vehicles where the entire vehicle is tested on a chassis dynamometer. On-road fuel economy can be measured; however, the wide variety of HD vehicles means lots of testing.

THE UNLEADED FUEL PARADOX

Taking lead out of gasoline had some good effects as in lower lead levels in the atmosphere and along roadways, and facilitating exhaust emission control by catalytic devices. However, taking lead out resulted not only in higher energy consumption in the refinery (lower yield), but a reformulation of gasoline toward use of more volatile and aromatic components. These were consequences that neither the government nor the car makers appeared to take into consideration in 1973, and resulted in what may be termed the unleaded fuel paradox. Cars are certified by EPA for sale in the U.S. based on unleaded reference fuels for emissions and durability testing[26]. These fuels, originally selected in the late sixties, were intended to typify fuels then sold commercially.

Volatility, Vapor Pressure and Volatile Organics - Volatile organic compounds (VOCs) are the major drawback to the future use of gasoline in the U.S. One reason gasoline is a good fuel is that its volatility makes it relatively easy to vaporize and mix with air and ignite, especially when cold. Thus, the volatility is adjusted regionally and seasonally following ASTM D 4814. The emissions certification gasoline is 9 psi vapor pressure and this was not atypical of the late sixties summer grade of gasolines. Except for California, which always enforced 9 psi summer vapor pressure, gasolines sold in the U.S. gradually increased in vapor pressure during the 1975-85 period to about 11.5 psi (summer grade). Although most cars ran satisfactorily, some of the vapor from the fuel tank and engine during diurnal heating and after engine shut-down was not collected by the activated carbon canisters. The increased vapor pressure was usually a result of blending butane[27] and more volatile aromatics into the fuel to achieve the octane specification. Butane has low primary atmospheric activity and is of little consequence in the photochemical reactions to form ozone. However, some of the evaporative emissions not collected include other VOCs, thought to be more photochemically reactive.

Thus, the fuel reformulation that occurred to produce increasingly larger quantities of unleaded fuel each year generally brought about more and different VOCs than those in the certification fuel. Their escape routes into the atmosphere became better understood as in refueling losses, the release of gasoline vapors equal in volume to the amount of fuel pumped into the tank. Running losses, vapors emitted from the vehicle while in operation other than tailpipe HC, were defined. Work on behalf of EPA[28] and unpublished results indicate that some of the best octane improving ingredients of gasoline do not oxidize easily by the exhaust catalyst. This has led to further speculation that perhaps some gasoline compositions resulted in tailpipe emissions with higher overall photochemical reactivity than was anticipated. Perhaps the way the unleaded gasoline was made could have had an unexpected impact on resulting air quality in terms of ozone.

In the mid-eighties, the use of various alcohols to supplement and extend gasoline were used, sometimes to the point that motorists began complaining of impaired driveability in terms of surge and stretchy-lean operation and lack of power from their cars during warm summer months. These symptoms of vapor lock resulted when blend vapor pressure of some summer gasoline grades reached as high as 13.5 psi. However, when the cars were fueled with a 9 psi gasoline, the cars ran fine.

The initial response to these complaints came from the government. More states began regulating vapor pressure and alcohol content while the EPA published rules[8] to limit summer vapor pressure to 10.5, 9.5 or 9 psi starting on May 16, 1989 (Stage I) and to 9, 7.8 or 7 psi on May 16, 1992 (Stage II) following regions as defined in ASTM D 4814. The California Air Resources Board (CARB) Stage I reformulated fuel requirements beginning January 1, 1992 is 7.8 instead of 9 psi. These changes will further impair refinery operation since a change in 1 psi has been estimated to lower the industry's gasoline capacity by about 2 percent[29].

Another response to the VOCs was to require additional Stage II refueling loss recovery as well as vehicle on-board recovery. The on-board recovery system would likely involve a much larger activated carbon canister than is currently used and could also collect vapors emitted by the fuel system and vehicle during operation (running loss). The on-board system is under review by the U.S. Department of Transportation (DOT) for possible vehicle fire safety implications.

A third response was the introduction by ARCO in September 1989 of EC-1, a reformulated fuel to replace leaded fuel in Southern California[6]. This fuel contains fewer of the lightest compounds such as butane, and only about one-third of the heavier forms of two classes of hydrocarbons, the aromatics and olefins. Compared with leaded regular, EC-1 has about one-third of the aromatics (including 50 percent of the benzene). It has reduced vapor pressure to 8 psi, 1 psi below the then existing California standard. As a result, tests show EC-1 to emit fewer reactive hydrocarbon VOCs[29]. Perhaps more than any other fuel's development in recent years, ARCO EC-1 reawakened the industry to the potential of further environmental improvements possible by reformulating gasoline. With the exception of MTBE, all these changes had been investigated in the sixties.

A fourth response to the VOCs has been to investigate ways to improve gasoline through reformulation and use of additives. Phase I of the Auto-Oil Program, started in 1989, makes use of an extensive matrix of fuel formulas, cars and measurements of regulated and unregulated emissions, including comprehensive HC speciation. Air quality modeling studies assess the anticipated improvement in air quality in terms of HC, ozone, CO, and NO_x. It is thought that one or more reformulated fuels will be identified for use in existing as well as new cars.

A fifth response to the VOCs concern was CARB regulations of September 1990 establishing standards based on non-methane organic gases (NMOG) adjusted for individual component reactivity as the method of expression for tailpipe and evaporative hydrocarbons. Using detailed HC speciation, and the relative reactivities on Table 1[30], the ozone making potential of gasoline HC can be computed. NMOG, adjusted for photochemical reactivity, is then a common basis for comparison and provides a "level playing field" for various reformulated gasolines, alcohols, and natural gas.

A sixth response to the VOCs was the continued pressure to use more and more alternative fuels instead of gasoline. The encouragement given FFVs and alcohol fuel, primarily M-85, by the Bush Administration in 1989[5] illustrates the extent of the concern. Other alternatives such as natural gas (compressed or liquid) and liquid petroleum gas gain credibility by replacing the volatiles which characterize today's gasoline and its atmospheric reactivity. Another response, but weakly voiced, is to use diesel fuel, which, because of its higher boiling range and general lack of volatile components that vaporize easily, would obviate the need for diurnal, hot engine soak, refueling, or vehicle running losses. However simple such a solution to the reactive VOCs from gasoline, the diesel has its own set of problems which will be discussed in more detail later.

After review of this litany of responses to gasoline VOCs, is there any doubt that this is the major problem faced by refiners? The paradox is that VOC emissions from gasoline powered vehicles were partially affected in quantity and type by an earlier reformulation, brought on when lead was removed.

Importance of Octane Improving Additives - Refinery yield, energy consumption and operational flexibility is enhanced to the extent the octane requirements can be met through additives. The search for substitutes for TEL and TML goes on. But the continued reluctance of CARB and EPA to consider metal containing additives such as methylcyclopentadienyl manganese tricarbonyl (MMT) has curtailed such research. The current concerns relate to increased engine-out particulates and health effects of MMT combustion products in the atmosphere. The uphill battle to obtain approval and acceptance of potent octane improvers such as MMT[31] has forced refiners to search for different and improved processing methods that achieve required octane at reduced vapor pressure. These interrelationships are very complex and are different for different refineries, crudes and product mixes. Complicating the issue further is the mandated addition of oxygen to the fuel by use of alcohols or ethers. For example, the increase in vapor pressure when using an alcohol to achieve the required oxygen content is only partially offset by the octane blending value of the alcohol. Table 2 lists the vapor pressure and octane blending values for several alcohols and ethers, classed as oxygenates.

Table 1 Maximum Incremental Reactivities of Selected VOCs[30]

Compound (VOC)	Mol. Wt.	Kinetic Reactivity	Mechanistic Reactivity	Overall Reactivity
		fraction VOC reacted	gm,O_3/gm VOC reacted	gm,O_3/gm VOC emitted
Carbon Monoxide	28	0.051	1.9	0.097
Methane	16	0.002	18.6	0.036
Ethane	30	0.24	8.3	0.49
Propane	44	0.43	4.3	1.00
N-Butane	58	0.58	2.8	1.85
N-Pentane	72	0.58	2.8	1.64
N-Hexane	86	0.69	2.2	1.50
N-Octane	114	0.83	1.2	0.98
N-Decane	147	0.91	0.8	0.77
Branched C-6 Alkanes	86	0.70	3.3	2.32
Branched C-8 Alkanes	114	0.85	2.0	1.71
Branched C-9 Alkanes	128	0.90	1.7	1.52
Ethene	28	0.93	16.4	15.25
Propene	42	1.00	12.7	12.66
1-Butene	56	1.00	12.0	11.98
Trans-2-Butene	56	1.00	13.4	13.35
C-8 Alkenes	112	1.00	6.2	6.16
Benzene	78	0.25	4.0	1.00
Toluene	92	0.73	7.3	5.31
P-Xylene	106	0.96	11.2	10.73
Tnalkyl Benzene	120	1.00	11.5	11.50
Acetylene	26	0.16	6.6	1.05
Methanol	32	0.19	7.4	1.40
Ethanol	46	0.51	4.4	2.25
t-Butanol	74	0.22	4.1	0.91
Formaldehyde	30	1.00	16.5	16.46
Acetaldehyde	44	1.00	11.6	11.55

Table 2 Some Characteristics of Oxygenates

	Alcohols		Ethers		
Characteristic	Methyl[1]	Ethyl	MTBE	ETBE	TAME
Oxygen Content, Weight %	50	35	18.2	15.7	15.7
Blending Vapor Pressure, psi	4.6	2.3	9	4	4.4
Octane Blending Value[2]					
Research, R	131.5	117.5	118	119	113.5
Motor, M	101.5	103	102	103	101.5
(R+M)/2	116.5	110.3	110	111	107.5
Volume % Required to Reach:					
2% O_2 by Weight	4	5.4	11	12.7	12.7
2.7% O_2 by Weight	5.4	7.3	15	17.1	17.2

(1) Must be blended with a cosolvent to be used as a gasoline additive

(2) Midrange octane blending values are approximate and vary with gasoline blendstock[20]

ENVIRONMENTAL LAWS

Since the first Federal tailpipe regulations for model year 1970 cars, the government has had its way with the car, truck and bus industry through progressively more stringent, technology forcing standards. Except for the lead phase-out and requirements to control gasoline vapor pressure, the government has not attempted to regulate the refiners in the same way as the manufacturers; that is until the Clean Air Act Amendments of 1990 (CAA'90)[32].

Federal Standards

Table 3 is a brief listing of standards for passenger car (LDV), defined as 6000 lbs. GVW or less, and light truck (LDT), vehicles with more than 6000 lbs. up to 8500 lbs. GVW. The CAA'90 requires further reductions for LDV by 1994 in NO_x of 0.4 g/mile (same as CARB 1989). The existing standards for LDTs are essentially halved and begin in 1995. Tier II of the CAA'90 provides for 50 percent reductions in CO, NMHC, and NO_x in 2004, unless EPA finds the new standard is not (1) necessary, (2) technologically feasible, or (3) cost effective.

Table 3 U.S. Vehicle Pollution Limits

	Grams/Mile		
	CO	HC	NO_x
	Existing Limits		
LDV 1989	3.4	0.41	1.0
LDT 1989	10.0	0.8	2.3
Evaporative	2.0 g/test		
Diesel LDV P.M.	0.2 g/mile		
	Clean Air Act Amendments of 1990		
		NMHC	
Tier I			
LDV 1994	3.4	0.25	0.4
LDT 1995	5.0	0.39	1.1
Tier II			
LDV 2004	1.7	0.125	0.2

Doubled Durability - The major problem expected with the Tier I regulations is the 100,000 mile durability demonstration, instead of the current 50,000 mile test. The manufacturer sees perhaps triple the recall liability, an ever present threat. In reality, cars usually are not recalled for repair beyond about 25,000 miles because by the time the "fix" is determined, the owners notified, the remaining life of the vehicle may not be worth the trouble, or at least this is one argument. If cars are not recalled for emissions repair after say 75,000 miles, the recall life is about three-fold over current practice. This forces manufacturers to not only perform much longer durability testing, but take all steps possible to guard against deterioration or failure on a vehicle that gets progressively less attention, less miles per year, and will be in service much longer with sometimes 2 or more owners.

Evaporative and Other LDV/LDT Needs - With regard to evaporative losses, the new Federal limits remain unchanged at 2 grams HC per test. This limit will be joined in 1996 by standards for vehicle running and refueling HC losses. Diesel cars currently must meet 0.2 g particulates per mile. Tier I diesel particulates for 1994 LDV will be 0.08 g/mile and 1995 LDT will be 0.12 g/mile.

Other on-highway LDV/T provisions in the CAA'90 relate to the allowance of other states to imitate the same CARB stringent car emissions standards. Also required will be expansion of inspection and maintenance requirements from 70 to 110 of the most affected areas, with improvements required in on-board diagnostics for emissions problems. LDV cold-start (20°F) CO standards begin in 1994 (Tier I) and become more stringent in 2001. In 1998, fleets of 10 or more centrally fueled vehicles in ozone non-attainment areas of over 250,000 population are encouraged to use alternative fuels such as methanol or natural gas in order to meet California low emission vehicle (LEV) standards.

HD Truck and Bus - The Federal emission standards for HD diesel engines were changed by the CAA'90. Listed in Table 4 are the current and future limits. The CAA'90 revised the urban bus limits, allowing the 0.1 g/hp-hr particulate limit to be delayed until 1993 with the change in 1994 to 0.05 g/hp-hr particulates or 2.5 g/hp-hr NO_x. Halving the already stringent 0.1/5 g/hp-hr particulates/NO_x limits requires buses to use fuels such as methanol or natural gas. A possible alternative, although one that may not be competitive, will be a low emission diesel fitted with a flow-through catalyst plus a highly efficient particulate trap. Another "surprise" was the further lowering of the HD NO_x standard to 4.0 g/hp-hr in 1998. By way of comment, some believe this to be the limit for in-cylinder NO_x reduction without increasing engine-out particulates.

Table 4 U.S. EPA Heavy-Duty Diesel Emission Regulations(1) (G/HP-HR)

	NO_x		PM
1988 (all)	10.7		0.6
1990 (all)	6.0		0.6
1991 (all)	5.0		0.25
1993 (urban buses)	5.0		0.10
1994 (trucks)	5.0		0.10
1994 (urban buses)	2.5	or	0.05
1998 (trucks)	4.0		0.10

(1) HC of 1.3 and CO of 15.5 g/hp-hr unchanged since 1988

Heavy-Duty Clean Fuel Vehicles - Section 245 of the CAA'90 requires vehicles above 8500 and up to 26,000 lbs. gross vehicle weight (GVW) to have, starting model year 1998, combined NO_x + NMHC of 3.15 g/hp-hr. This is equivalent to 50 percent of the 1994 HD diesel limits. This weight category includes two-axle trucks and small buses. This, plus the city bus limits of 2.5 NO_x or 0.05 particulates, portends perhaps even more stringent regulations for trucks and buses used in and around cities. The advent of a non-methane standard for HD engine HC encourages natural gas since the HC part of the regulation will not be as technology forcing. It is speculated that once the 1994 limit of 0.05 grams particulate or 2.5 grams NO_x is demonstrated for urban buses, other diesel powered vehicles used in sensitive areas that continue to have pollution problems, will have such regulations. Perhaps there will be a 0.05 g/hp-hr organic particulate standard for all HD by 2000. This will remain to be seen and may depend on if and when efficient and durable aftertreatment systems are developed.

Potential Methane Standard - The adoption of a methane HC standard is of interest if:

(1) international agreements on reduction of greenhouse gases is reached that would require control of methane emissions,

(2) natural gas HD vehicles become popular, and

(3) many cars and light trucks are converted to natural gas fuel.

The likelihood of the above is not completely clear, yet are distinct possibilities making NMHC standards that intentionally delete methane as another potential paradox. Some currently evaluated trucks converted to natural gas have very high hydrocarbons. Assuming 90 percent of the exhaust HC is methane, the NMHC standard will allow the engine to emit up to 13 grams HC per bhp-hr. Discounting unburned lubricant, 11.7 grams would be methane and the NMHC would be 1.3 grams, the diesel standard. In the event a methane standard is promulgated, there is some possibility of catalysts being developed for oxidation of methane. However, applications to vehicles and durability remains to be demonstrated. Thus, a NMHC standard encourages waste of fuel as well as permits greater emission of a greenhouse gas that has ten times the life of CO_2 and is 30 times more effective than CO_2[13].

Off-Road and Locomotives - The CAA'90 also calls for regulation of non-road engines and vehicles greater than 175 hp starting in 3 years. Locomotives are to be regulated in 5 years. Regulations for off-road and locomotives are to be based on the greatest degree of emissions reduction achievable.

CARB Standards

Continuing their self-appointed role as leader in vehicle air pollution regulations, the CARB adopted standards in 1990 for model years out to 2003[33]. Table 5 lists four new standards for transitional low emission vehicle (TLEV), low emission vehicle (LEV), ultra low emission vehicle (ULEV), and zero emission vehicle (ZEV) beginning with the 1993 model year.

Table 5 CARB Car Standards, Gasoline Fueled, g/mile (50,000 mile)

	NMHC	CO	NO_X	PM(2)	
1989	0.39	3.4	0.4	0.08	
	NMOG(1)				HCHO(3)
1993	0.25	3.4	0.4	0.08	0.015
TLEV	0.125	3.4	0.4	0.08	0.015
LEV	0.075	3.4	0.2	0.08	0.015
ULEV	0.040	1.7	0.2	0.04	0.008
ZEV	0	0	0	0	0

(1)Non-methane organic gases (NMOG) includes HC, oxygenated HC such as alcohols, aldehydes, ketones, and ethers and would be adjusted for clean fuels reactivity
(2)PM: diesel vehicles only
(3)HCHO: methanol (M-85) vehicles only

Hydrocarbons Redefined as NMOG Adjusted for Reactivity

Future hydrocarbon limits subtract the methane content, a negligible emission from the micro-environmental viewpoint for gasoline powered cars. CARB redefines hydrocarbons for post-1993 low emission vehicles as non-methane organic gases - NMOG. NMOG includes the usual HC plus the oxygenated HC such as alcohols, aldehydes, ketones, and ethers. The important twist in the use of NMOG is the requirement to analyze the HC for its constituents and then to apply their relative reactivities to determine ozone-forming potential. California has based their standards on currently specified certification gasoline, described in Table 6[26]. Through the adjustment of the various HC components for their relative reactivity, a basis for comparison with other fuels such as M-85 or natural gas or with various reformulated gasolines is provided.

Table 6 Specification of Unleaded Gasoline Used in U.S. Emissions Certification [26]

Item	ASTM	Unleaded
Octane, research, minimum	D2699	93
Sensitivity, minimum		7.5
Lead (organic), grams/U.S. gallon		0.00-0.05
Distillation Range:		
IBP[1], °F	D86	75-95
10 percent point, °F	D86	120-135
50 percent point, °F,	D86	200-230
90 percent point, °F	D86	300-325
EP, °F (maximum)	D86	415
Sulfur, weight percent, maximum	D1266	0.10
Phosphorous, grams/U.S. gallon, max		0.005
RVP[2,3] pounds per square inch	D323	8.7-9.2
Hydrocarbon Composition:		
Olefins, percent, maximum	D1319	10
Aromatics, percent, maximum	D1319	35
Saturates	D1319	([4])

[1]For testing at altitudes above 1,219 m (4,000 ft) the specified range is 75-105.
[2]For testing which is unrelated to evaporative emission control, the specified range is 8.0-9.2.
[3]For testing at altitudes above 1,219 m (4,000 ft) the specified range is 7.9-9.2.
[4]Remainder.

For example, using certification gasoline, on which the 0.075 g/mile LEV NMOG is based, an FFV burning M-85 could be allowed to meet a higher standard, perhaps 0.1 to 0.15 g NMOG/mile, depending on the specific FFV emissions. Instead of M-85, maybe a modified car operating on a reformulated gasoline could achieve reduced reactivity adjusted NMOG. This is one important outcome expected to be identified in the Auto-Oil Program.

ULEV-NMOG Limit Achieved - Alternative fuels, such as M-85 or natural gas, may not be necessary for manufacturers to meet the ULEV limits of 0.04 g NMOG/mile. It has been shown, at least to the CARB satisfaction, that cars have the potential for achieving 0.04 g NMOG/mile operating on certification gasoline. This feat was accomplished by Southwest Research Institute's Emissions Research Department in

1990[34]. By injecting air into an electrically heated catalyst during the first 60-80 seconds of cold-start, HC were effectively controlled until the main exhaust catalyst reached operating temperature.

For many years, it has been known that the cold start emissions during the first few minutes of the Federal Transient Procedure (FTP) after cold start, remained uncontrolled and could contribute more HC than the remainder of the test. To get the engine to start, most fuel control systems go "open-loop" during which time a richer than stoichiometric mixture is provided in order for the engine to start and run. As better catalysts and controls were developed, the open loop time has been steadily reduced to about 1 minute. Figure 5 illustrates the cold start control of HC from a 1990 catalyst equipped car.

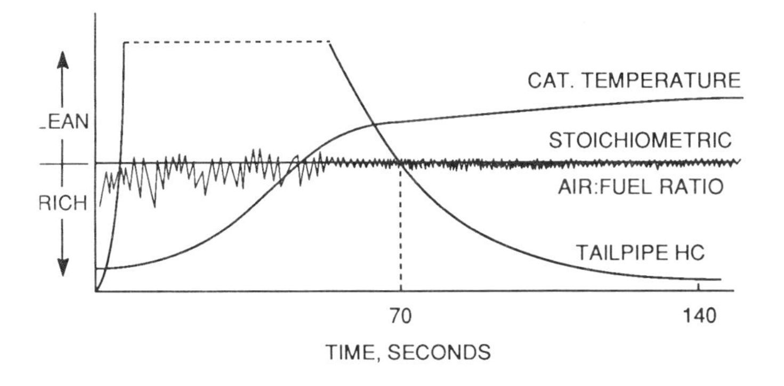

Figure 5 Typical 1990 Model FTP Cold Start Catalyst and Hydrocarbon Behavior

In the SwRI studies, it was found that by preheating a small metallic oxidation catalyst located in front of the main catalyst, and controlling the injection rate of supplemental air for combustion, the excess hydrocarbons can be catalytically burned. In September 1990, two cars achieved 0.03 g NMOG/mile in laboratory tests at SwRI and at CARB[35].

Although there are many questions remaining on practicality, durability and customer acceptance of the preheat delay time plus battery requirements, a method has been shown to achieve the ULEV NMOG using certification gasoline and without necessarily resorting to an alternative fuel. This development also re-identifies the one remaining tailpipe emissions problem for gasoline cars, the cold-start, an old problem that has progressively been reduced by the manufacturer. There are many alternative ways to reduce the cold-start HC, such as better pre-cylinder air-fuel mixture preparation through improved injectors and location, mechanical and thermal vaporizers, and relocated oxidation catalysts with even lower light-off temperatures.

<u>CARB Fleet Avg. NMOG Schedule</u> - Table 7 shows percent sales of cars in California by model year to achieve the fleet average NMOG. This is an example since the manufacturer can use other introduction dates and even skip TLEV cars, as long as the fleet average is attained. LEVs are expected to be the dominant car in the product mix between 1988 to 2003. Some hybrid engine-electric and pure electric cars such as the GM Impact[36] could start selling before the year 2000 as a part of the strategy to meet the fleet average NMOG.

Fuels

Why are so many of the major air pollution areas of the U.S. still "out of attainment" for ozone and CO? This recurring question is central to the CAA'90 and will be for the decade of the nineties and perhaps beyond. "Out of attainment" means the locale still has atmospheric levels that exceed the National Ambient Air Quality Standards (NAAQS) for one or more primary pollutants. Local, regional, and state implementation plans were intended to achieve such limits under deadlines imposed by the EPA by a mix of controls on stationary and mobile sources. Why did such plans, models, and controls fall short in so many cities on ozone and CO? The NAAQS for ozone is 235 micrograms per cubic meter (0.12 ppm) on an annual basis while CO is 10 milligrams per cubic meter (9 ppm) on an 8-hour basis[37].

<u>Federal Regulations, Gasoline</u> - The CAA'90 has two major fuel requirements; a reformulated gasoline for ozone control and a gasoline with increased oxygen for CO control. A reformulated gasoline is to be sold starting in 1995 in the nine worst ozone non-attainment cities in the U.S. These cities are Baltimore, Chicago, Hartford, Houston, Los Angeles, Milwaukee, New York, Philadelphia, and San Diego. They represent about 30 percent of the U.S. population and account for about 25 percent of gasoline sales. The reformulated fuel must contain 2 percent by weight oxygen, limit aromatics to 25 percent by volume, and limit benzene to 1 percent by volume. The fuel, as well as all gasoline sold in the U.S., must contain an additive to prevent injector and intake system deposits. In addition to these "recipe" specifications, the fuel must be qualified as to its effect on emissions of VOCs and toxics. A 15 percent reduction in VOCs and a 15 percent reduction in toxics is required by the reformulated fuel in 1995 with a target of 25 percent in each category in 2000. The EPA is expected to issue procedures by which these reductions are to be demonstrated and the reformulated fuels are qualified for sale. The basis for reduction will be a 1990 baseline fuel in a 1990 model car.

To reduce CO during winter months in 41 cities with CO problems, i.e., greater than 9.5 ppm CO ambient levels, the oxygen content in gasoline is to be increased to 2.7 percent by weight. This fuel is to be available in 1992, but may be delayed for up to 2 years by EPA if a shortage of oxygenates exists in a given area. This regulation makes use of the favorable Denver Front Range (Ft. Collins to Colorado Springs) experience of January 1 to February 28, 1988. During this time, gasoline containing 1.5 percent oxygen reduced ambient CO by 9 percent, the intended result[38]. Since then, the requirement has been to add 2 percent oxygen by weight during November 1 to February 28 each winter. During the first year, MTBE, rather than methanol, was used in 94 percent of the gasoline because of its superior blending characteristics, vapor pressure, and cost.

One way to achieve 2.7 percent O_2 by weight, required by the CAA'90, is to blend in 15 percent by volume of MTBE. Other oxygen containing chemicals, such as ETBE, TAME, ethanol or methanol may also be used to increase fuel oxygen content. The use of oxygenates can result in increased octane of the blend depending on the oxygenate and blendstock (Table 2). Perhaps this will allow components to be used in such formulas that result in fewer reactive NMOGs in the exhaust and from evaporation. Alcohols tend to increase blend vapor pressure through formation of azeotropes. This is expected to compensate for changes in the volatile components, as in fewer volatile aromatics.

Table 7 CARB Fleet Average NMOG Car Schedule (percent sales in California)

Model Year	NMHC 0.39	NMOG 0.25	TLEV 0.125	LEV 0.075	ULEV 0.040	ZEV 0	Fleet Avg
1994	10	80	10				0.250
5		85	15				0.231
6		80	20				0.225
7		73		25	2		0.202
8		48		48	2	2	0.157
9		23		73	2	2	0.113
2000				96	2	2	0.073
1				90	5	5	0.070
2				85	10	5	0.068
3				75	15	10	0.062

Reformulated Gasoline Observations - Technical Bulletin No. 1 of the Auto/Oil Air Quality Improvement Program[39] provided the initial results and effects of fuels on emissions in two fleets of vehicles: current (1989) and older (1983-85) models. This report was widely quoted as to the varying effect fuel composition changes had on regulated emissions of HC, CO and NO_x. Emissions from some cars would increase, some would decrease and the effect was termed a "mixed bag." The only consistent effect was that of fuel oxygen content on CO production.

The real significance in reformulation lies with the newly regulated species, the reactive HC, the VOCs and the toxics. It would be nice if a fuel change resulted in less NO_x. Just as important, perhaps even more important, would be reduced reactive HC. The possibilities of this should be available soon from air quality modeling of the extensive tailpipe and evaporative HC speciation. For reformulated fuel to compete with alternatives, the potential to form ozone is crucial. In the meantime, General Motors recommended, effective March 18, 1991, the use of reformulated gasoline in all its existing and future model cars and trucks. Preferring not to wait until 1995 for the CAA'90 requirements, the 1992 GM Owners Manuals will contain the reformulated fuels recommendation[40].

The Auto-Oil results to date confirm that learned at SwRI in independent studies. Observations from many fuel-vehicle matrix projects may in general be summarized as follows.

Fuel Modification	Emission Effect
Decrease volatility	Decrease exhaust and evaporative HC emissions
Decrease aromatic content	Decrease exhaust and evaporative aromatic emissions
Add oxygenates	Decrease CO Increase iso-butylene (MTBE,ETBE) Increase formaldehyde (MTBE) Increase acetaldehyde (ETBE, ethanol) Reduce fuel economy
Decrease olefins	Decrease unburned fuel olefins

Except for the effect of oxygenates, there are exceptions to all of the above in that HC, CO and NO_x can vary up to three-fold with fuel modification. Another observation was that high octane components in the fuel, such as iso-octane, toluene, xylenes, ethylbenzene, and benzene, are the most difficult to remove with catalytic aftertreatment while 1,3-butadiene in the exhaust is easy to remove during the cold-start with an air injected, electrically heated catalyst[34,35].

Federal Regulations, Diesel Fuel - The importance of diesel fuel sulfur content on sulfuric acid particulate emissions of the diesel engine has been well established[41]. Figure 6 illustrates the importance of reducing fuel sulfur content on achieving the 0.1 g/hp-hr standard, as well as allowing use of a flow-through oxidation catalyst. This type of aftertreatment is expected to be sufficient for most engines to meet the 1994 regulation. In an unprecedented action, the Engine Manufacturers Association (EMA) and American Petroleum Institute (API) plus two other industry groups, held a series of meetings on the need to reduce sulfur and aromatics in diesel fuel. A joint report[43] issued in July 1988 recommended that EPA require all on-highway type 2-D fuel be reduced in sulfur content to 0.05 percent by weight maximum effective October 1, 1993. Furthermore, it was recommended that aromatics be controlled by requiring such fuels to have a minimum cetane index of 40, the current minimum cetane number according to ASTM D 975-81 for type 2-D fuel. The relationship between calculated cetane index and engine cetane number is very close, unless cetane improving additives are used.

On August 21, 1990, EPA issued a rule[44], since confirmed by the CAA'90, that endorsed the industry recommendations and added an alternative 35 percent aromatics maximum[44]. This fuel is expected to be about 2 to 3 cents per gallon more at the pump overall. The price increase will depend on the size and type of refinery, crude oil available, and product mix in the refinery.

CARB Gasoline Regulations - In September 1990, the CARB issued their Phase I regulations for gasoline effective January 1, 1992. The summer vapor pressure (April - October 13) is lowered from 9 to 7.8 psi, with limits of 0.05 g/gal lead and 0.005 g/gal phosphorous. The lead and phosphorous limits are to reduce deterioration of oxygen sensor and catalyst life. The gasoline must also control port fuel injector deposits to a maximum flow loss of 5 percent, using the Coordinating Research Council (CRC) 15 minute run - 45 minute off (soak)

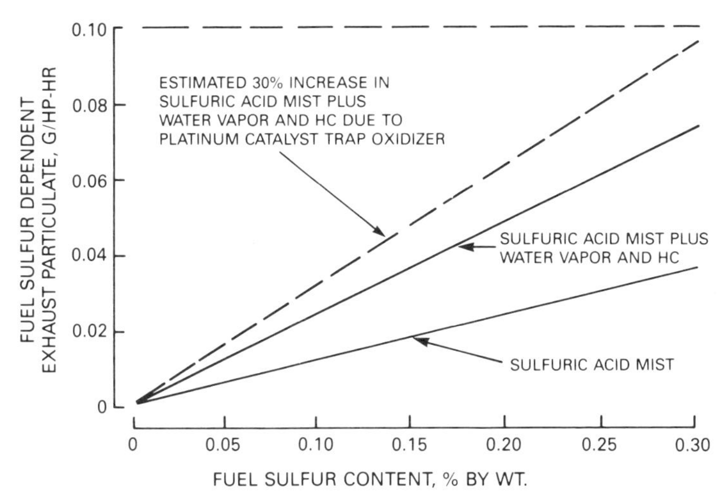

Figure 6 Collectible Fuel Sulfur Dependent Particulate as a Function of Fuel Sulfur Content

Figure 7 Intake Port of Fuel Injected Gasoline Engine[48]

cycle car test[45]. A Shell bench test for port fuel injector deposits holds promise for predicting stay-clean and clean-up performance of additive-treated gasolines. Its development at SwRI has been supported by CRC and industry. The method is very similar to that described in reference 46.

Intake valve deposits must also be controlled to a maximum of 100 mg/valve average using the 10,000 mile SwRI-BMW intake valve deposit (IVD) test[47]. In order to achieve vaporization of the injected fuel spray before ignition, the fuel is intentionally directed onto the hot underside of the intake valve. The valve surface temperature can approach 600°F. If the fuel decomposes and leaves a residue in the intake system, it can, over time, cause customer complaint during the crucial cold-start, drive-away type of vehicle operation. The fuel, during cold-start, tends to be adsorbed by the deposits during idle and initial acceleration. In this way, the engine can operate so lean that the engine stalls, stumbles on acceleration, lacks power, and has a rough idle. As soon as the engine reaches operating temperature, adsorption-desorption equilibrium is achieved. Figure 7 is a cutaway of an intake port showing the fuel injector and intake valve relative to the fuel injector[48].

There were sufficient complaints about Bavarian Motor Works (BMW) car driveability, due to IVD, that work to develop a fuel qualification procedure was begun in 1986. Like GM, BMW went public with their problem and convinced the refiners to qualify their gasolines using a 10,000-mile car test based primarily on intake valve deposits and secondarily on driveability and other car performance factors. Pass/fail limits for the gasoline of <100 milligrams (mg) average deposit acceptable for unlimited use and 250 mg for 50,000 miles use were established. Figure 8 shows the typical deposits for the pass/fail acceptance criteria.

At the February 28, 1991 CRC Intake Valve Deposit Group meeting, Ford Motor Company reported that 97 percent of 1990 model year cars and 100 percent of 1990 model year light trucks were designed for regular grade unleaded gasoline, (R+M)/2 of 87. Ford further reported that the fuel quality statement in the Owners Guide will be revised to state:

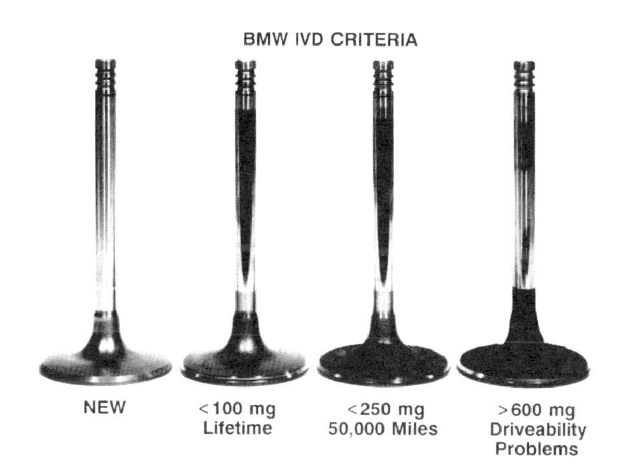

Figure 8 Pass/Fail Criteria for Gasoline Qualification 10,000 Mile BMW-SwRI Intake Valve Deposits[48]

- "All grades of unleaded gasoline should contain an effective intake system deposit control additive
- Do not use a higher octane grade unleaded gasoline than recommended in the Owners Guide"

By way of explanation, Ford said that premium grade unleaded fuels in vehicles designed for regular grade unleaded may aggravate driveability symptoms. They requested a gasoline deposit performance test procedure be included in SAE J312/ASTM D 4814 requirements[49]. Fuel additives, as described for example by reference 50, will play a major role in maintaining excellent driveability of cars through improved gasoline performance[51].

Gasoline will be further modified, perhaps reformulated, following CARB Phase II rule making expected in September 1991. Under consideration will be new specifications on aromatic content, vapor pressure, olefins, distillation, benzene, oxygen content, and ozone forming toxic potential. Some of the expected changes are in the same direction as the CAA'90 fuel for the nine worst O_3 non-attainment areas and would take effect perhaps in 1994, 1995 at the latest.

CARB Section 2345, Substitutes for Clean Fuels, specifies what a gasoline must achieve if it is to be a substitute for a "clean fuel." Primary "clean fuel" examples are methanol (M-85 or M-100), ethanol (E-85), compressed/liquid natural gas (C/LNG), or liquid petroleum gas (LPG). The rules are effective from 1994 to 1996 in the Los Angeles Basin for large refineries (>50,000 BPSD) and in 1997 statewide, all suppliers.

The substitute fuel, termed "deeply reformulated" by the author, must demonstrate emissions equal to or less than a LEV operating on one of the primary "clean fuels." Reactivity adjusted tailpipe and evaporative NMOG, NO_x, CO, and the aggregate toxic-weighted benzene, 1,3-butadiene, formaldehyde, and acetaldehyde are the emissions on which the comparison will be made. Particulates will also be used for comparison if diesel fuel is considered as a substitute. Qualification of a "deeply reformulated" fuel against a low emission FFV burning M-85 is a major test for reformulated gasoline. The NMOG, adjusted for atmospheric reactivity, is intended to allow direct comparison.

CARB Diesel Fuel Requirements - On November 22, 1988, CARB adopted regulations for on-highway type 2-D fuel sold beginning October 1, 1993. The low sulfur limit of 0.05 percent by weight is the same as EPA[47]. However, CARB specifies that the fuel have less than 10 percent aromatic content for large refineries of greater than 50,000 barrels per stream day. Figure 9 illustrates the aromatic content of some 1393 samples collected over a 5-year period. As may be seen, few diesel fuels were sold below 20 percent aromatics, with the average of about 35 percent evident.

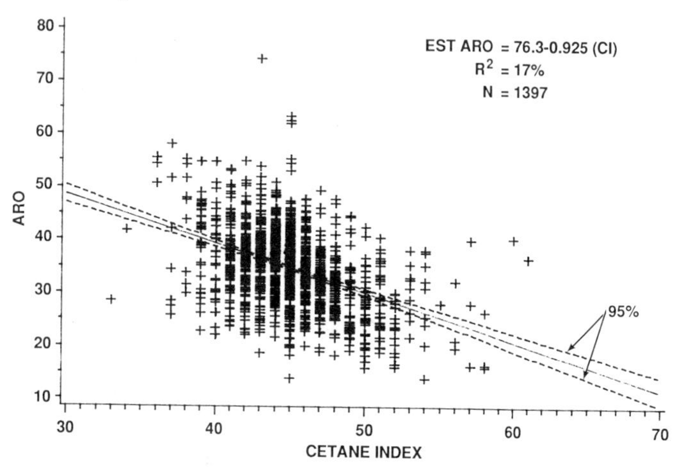

Figure 9 Aromatics and Calculated Cetane Index for 1397 Samples Summer 1983 to Winter 1988 Diesel Fuel Survey[48]

In the CRC VE-1 project, it was found that aromatic content had little effect on particulates, but an increased effect on NO_x for a 1991 prototype diesel engine (Detroit Diesel Corporation Series 60)[52]. Both particulates and NO_x from then current technology engines were sensitive to aromatics as shown in Figure 10[48,53]. To apparently achieve both low particulates and NO_x from future diesel engine fuels, a 10 percent aromatics content was specified; even though none are made at that level and the cost to do so in refinery energy and yield may prove very expensive, even prohibitive for some suppliers. One estimate indicated that making 10 percent aromatic fuel could increase the price by as much as 20 cents per gallon for those refiners that elect to do so.

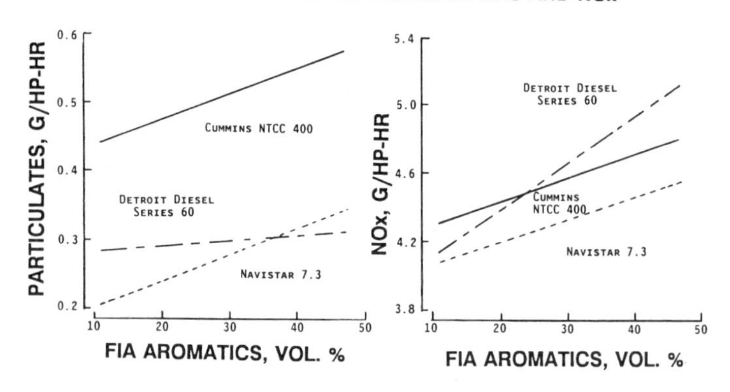

Figure 10 Effect of Aromatics on Particulates and NO_x from Three HD Diesel Engines by EPA Transient Test[48,53]

CARB, however, leaves a way out for the refiner in that he can, if he wishes, offer a reformulated type 2-D diesel fuel if it gives emissions equal to or less than would be obtained from a 10 percent aromatics fuel[54]. The test procedure is basically the same as that used in the CRC VE-1 project, namely the Detroit Diesel Corporation direct injected Series 60, an 11.1 liter, 4-stroke cycle turbocharged and aftercooled engine of 330 hp at 1800 rpm. The procedure involves replicate transient cycle tests using the EPA HD diesel engine certification procedure, in which the 10 percent aromatic reference fuel is run back-to-back with the candidate fuel. The CRC VE-1 work was continued by the Western States Petroleum Association to look at the effect of cetane level from natural and chemical additive means to achieve particulates and NO_x similar to that achieved with 10 percent aromatic content. Table 8, calculated from reference 55 data, indicates that the natural cetane could substitute for reduced aromatics more effectively than a cetane improver such as 2-ethylhexyl nitrate.

Table 8 Substitution of Cetane for Aromatics(1)

	Natural Cetane	Improved Cetane(2)	Aromatics %FIA
Particulates:	52	55	25
	50	53	20
	49	51	15
NO_x:	65	88	25
	58	70	20
	52	57	15

(1) Examples calculated to achieve emissions equivalent to 48 CN, 10% aromatic reference fuel
(2) Improved with 2-ethylhexyl nitrate. 45 CN national average used as baseline fuel

For example, to make a diesel fuel with 20 percent aromatics equivalent in particulate emissions to the 10 percent aromatics, 48 cetane CARB reference fuel requires a fuel with a natural 50 cetane or a cetane improved fuel by chemical additive of 53. This is not so difficult to achieve considering national average diesel fuel is 45 cetane number. However, in the case of NO_x, the 20 percent aromatics fuel will require either a 58 natural cetane or a cetane by chemical additive of 70. These calculations indicate natural cetane is a more effective substitute for aromatics on particulate and NO_x than 2-ethylhexyl nitrate. To produce equivalent NO_x to that from the 10 percent aromatic reference fuel will require substantial cetane improvement for the examples shown. It is believed that for most refineries, production of diesel fuel containing less than 20 percent aromatics will require major refinery change and be quite expensive.

The above, plus other possibilities of diesel fuel reformulation for California, are being investigated by several oil refiners to establish sensitivity of emissions to various reformulated, low sulfur, diesel fuels. So far, the possibilities of reformulation and its impact on price and availability of diesel fuel in California are largely unknown. Reference 54 provides details on the CARB procedure used for a reformulated type 2-D fuel qualification as alternative to the 10 percent aromatics, 0.05 percent sulfur by weight fuel.

Alternative Fuels

The major force driving regulations that encourage alternative fuels is environmental improvement. Other reasons such as less reliance on foreign crude oil or global warming have been used but the concern over VOCs from gasoline and organic particulates from diesel and their toxicity appear to be the root causes. The subject is of immense importance and has far reaching implications beyond the scope of this lecture. A few points need to be made in summary, however.

Methanol - Methanol promises to be a convenient way to get remotely produced natural gas to the U.S. market. From an energy-efficiency standpoint, it becomes attractive when the engine is designed to run with the fuel undiluted, as M-100, where compression ratio is increased and optimized. Mixtures of methanol and gasoline, intended to bridge the transition from gasoline to methanol, are not more fuel efficient and are questionable so long as gasoline is available and inexpensive. Environmental improvements seem a paradox. Everyone is aware of the claims for lower NO_x, volatiles, and reactivity adjusted NMOG. The major reactive component in M-85 fueled FFV exhaust is formaldehyde. According to Table 1, formaldehyde is 457 times more reactive in forming ozone than methane. Although some current technology catalysts are capable of efficient conversion of formaldehyde when new[56], such efficiencies have yet to be demonstrated with extended mileage. As shown in Table 5, the 50,000 mile CARB formaldehyde standard is 0.015 g/mile (TLEV and LEV) and 0.008 g/mile (ULEV).

Natural Gas - If methanol is a liquid form of natural gas, perhaps it would be more energy efficient to just liquify the gas and ship it to market in LNG tankers rather than make it into methanol. Then, it could be used in liquid or compressed state in vehicles. Aside from storage density and other concerns, natural gas makes sense if gasoline and diesel fuel are to be replaced. Again, to achieve the efficiency of this high octane fuel, the engine needs to take full advantage of the special features of natural gas while designing for low NO_x, the major problem expected. Depending on global warming requirements, a methane standard may be the only way to reduce methane emitted into the atmosphere from the engine or leaks. Regardless of the price and abundance of natural gas, its adverse potential as a greenhouse gas means that care must be taken in converting car and heavy duty engines.

Ethanol - Its major use remains as a 10 percent mix in gasoline as Gasohol. Ethanol and other alcohols such as tertyl butyl alcohol remain as gasoline fuel supplements and are used more or less depending on availability and price. It remains to be seen the extent to which ethanol will be used to achieve the 2 or 2.7 percent by weight oxygen requirements of future gasolines.

Ethers - Considered by some refiners as the most effective and easily blended oxygenates, methyl tertiary butyl ether (MTBE), ethyl tertiary butyl ether (ETBE), and tertiary amy methyl ether (TAME) are popularly used to add oxygen to gasoline. Beyond use in winter months in Colorado to reduce CO, these oxygenates will find increasing use in CAA'90 requirements for gasoline in the 9 worst ozone cities and in the 41 highest CO cities. Other uses for oxygenates are known. For example, two-stroke cycle gasoline powered motorcycles in Taiwan operate on fuel containing 2 percent oxygen by weight for more complete combustion and less CO. Additive treatment through oxygenates has great possibilities for improved, more complete combustion and fewer emissions. Similar additives may likewise help to reduce emissions from diesel engines. The regulations seem to encourage deposit control and combustion improving type chemicals as long as they are not metal based.

ACTION ITEMS

More than a wish list, the following action items will serve as discussion and summary.

1. Gasoline Octane - The importance of this single fuel property on refinery and vehicle efficiency and energy consumption cannot be minimized. For example, the use of metal-containing octane booster additives such as MMT needs to be resolved as to their acceptance by EPA and CARB.
2. Optimum Octane & Energy - The optimum octane number of gasoline for overall energy efficiency is a delicate balance between petroleum processing, environmental requirements, and engine thermal efficiency. A continuing study and optimization of the overall best octane quality for energy and efficiency is needed. It should make use of recent findings by the Auto-Oil Program, new refinery technology, and fuel additives development. The results of this study would be important in the national energy strategy.
3. Global Catastrophe - In any response to future agreements to reduce greenhouse gases because of global warming fears, standards for methane and CO_2 emissions are options as well as encouragement to conserve and reduce energy consumption.
4. CAFE - Energy conservation in transportation means CAFE to many while it is efficiency that is important. There is a need to consider innovative means of expressing and achieving economy beyond simply percentage roll-back or arbitrary miles per gallon levels. What, if anything, can be done to increase truck and bus fuel efficiency and economy? How to prepare CAFE rules that are appropriate for alternative fuels for FFVs, natural gas powered vehicles and electric or hybrid-electric cars? CAFE, to be meaningful, must reflect safety, customer, and society needs, as well as conservation of energy.

5. HD Particulates - The flow-through oxidation catalyst appears the technology of choice for HD diesel truck engines to meet stringent 1994 particulate standards. Long life of the catalyst, up to 290,000 miles for the largest trucks, will require identification of lubricants that are compatible with the catalyst operation, as well as work well in low oil consumption engines of the future.

6. Alternative Fuels - Lower particulate emissions, such as the 0.05 g/hp-hr required of city buses in 1994, will require alternative fuels such as methanol and natural gas or extensive exhaust aftertreatment. It is speculated that a flow-through catalyst plus a system of dual wall flow type ceramic filters, located as far from the engine as possible, to promote condensation of organics and aerosols on the collected carbon, might make the advanced technology diesel engine burning diesel fuel acceptable in environmentally sensitive applications[57].

7. Particulate Control - The flow-through catalyst plus cool exhaust wall filter is expected to remove nearly all particulates and organics. To what extent this technology can be applied to the diesel powered car is uncertain, since NO_x is an emission that must also be controlled. However, if the exhaust of the direct injected diesel car engine was benign, then its greatly reduced exhaust and evaporative VOCs, relative to gasoline, could be taken advantage of. Other well understood improvements in fuel economy and lower CO_2 emissions (global warming) could be realized.

8. Reformulated Fuels - Reformulated gasoline and diesel fuel cry-out for definition beyond that afforded by the CAA'90 or CARB regulations. There is much confusion over just what a reformulated fuel is, much less how to demonstrate and qualify it for use. The EPA is expected to publish procedures by which to demonstrate effectiveness of injector and intake system deposit control as well as the required 15 percent reductions in VOCs and toxics. Much more appears needed to define NMOG reactivity as required by the CARB for broad acceptance by industry.

The fundamental question of why so many cities failed to attain ozone and CO NAAQSs remains unanswered. All that is certain is the industry has another round of vehicle standards and recipes for reformulated fuels. Ten years from now, will these areas still be "out-of-attainment" and continue to elude our best efforts at control? It would be instructive to the future to understand the effect various factors had in the past such as deficiencies in modeling, emission factors, vehicle usage, vehicle miles traveled, maintenance, tampering, fuel quality, atmospheric interactions, hydrocarbon reactivity, evaporative losses, and other variables such as unregulated sources and pollutants in the mobile source pollution estimates. Perhaps such a look back will be valuable in the design of control programs and assessing progress toward environmentally acceptable and efficient transportation of the future.

LIST OF REFERENCES

[1] Clarke, P.J., Gerrard, J.E., Skarstrom, C.W., Vardi, J., and Wade, D.T., "An Adsorption-Regeneration Approach to the Problem of Evaporative Control," SAE Paper 670127 presented at the 1967 Winter SAE International Congress, Cobo Hall, Detroit, Michigan.

[2] 40 CFR, Part 86, Subpart A, para. 86.130-78, p. 503, Protection of the Environment, July 1, 1985.

[3] H.H. Kehrl, General Motors Vice Chairman, letter to Chief Officers of twenty-two petroleum companies, November 18, 1985.

[4] *Chicago Tribune*, "Oil, Auto Industries to Test Reformulated Gasoline," October 18, 1989.

[5] *Automotive News*, "Industry Awaits Details of Air Plan," p.49, June 19, 1989.

[6] *New York Times*, "ARCO Offers New Gasoline to Cut Up to 15% of Old Cars Pollution," August 15, 1989.

[7] Colluci, J.M., "Automotive Fuels for the 1990's - Challenges and Opportunities," *General Motors Research Laboratory Report GMR-6589, F&L-882*, March 9, 1989.

[8] 40 CFR, Part 80, EPA Volatility Regulations for Gasoline and Alcohol Blends Sold in Calendar Years 1989 and Beyond; Final Rule, March 22, 1989.

[9] "Alternative Transportation Fuels: No Simple Solutions Available," *Energy Focus*, Society of Petroleum Engineers, April 1990.

[10] *Bus Ride*, "A Farewell to Responsibility," May 1990.

[11] *National Wildlife Federations Conservation 90*, Vol 8, No 6, August 17, 1990.

[12] *The Oil Daily*, "Negotiations, Estimations and Rhetoric Leave Greenhouse Gas Issue Clouded," p. 5, February 12, 1991.

[13] Springer, K.J., "Global What? Control Possibilities of CO_2 and Other Greenhouse Gases," ASME Paper 91-ICE-10 presented at the Energy-Sources Technology Conference and Exhibition, New Orleans, Louisiana, January 20-24, 1991, and accepted for publication in the ASME Journal of Engineering for Gas Turbines and Power, July 1991 issue.

[14] Hammerle, R.H., Shiller, J.W., Schwarz, M.J., "Global Climate Change," ASME Paper 91-ICE-12 presented at the Energy-Sources Technology Conference and Exhibition, New Orleans, Louisiana, January 20-24-1991.

[15] Amann, C.A., "The Passenger Car and the Greenhouse Effect," SAE Paper 902099 presented at the SAE International Fuels and Lubricants Meeting and Exposition, Tulsa, Oklahoma, October 22-25, 1990.

[16] NRC, "1990 Fuels to Drive Our Future," *National Research Council Report*, National Academy Press, Washington, D.C.

[17] Amann, C.A., "The Automotive Spark-Ignition Engine - An Historical Perspective," ASME Paper, ICE-Vol. 8, pp. 33-45, presented at the Eleventh Annual Fall Technical Conference of the ASME Internal Combustion Engine Division, Dearborn, Michigan, October 15-18, 1989.

[18] Carroll, J.N., Ullman, T.L., and Winsor, R.E., "Emission Comparison of DDC 6V-92TA on Alcohol Fuels," SAE Paper 902234 presented at the SAE Truck and Bus Meeting and Exposition, Detroit, Michigan, October 29 - November 1, 1990.

[19] *Automotive News*, "Environmental Groups to Seek Boost in CAFE," p. 9, December 24, 1990.

[20] CONCAWE, Ad Hoc Group Automotive Emissions Fuel Characteristics. "Assessment of the Energy Balances and Economic Consequences of the Reduction and Elimination of Lead in Gasoline." R.Kahsnitz *et.al.,* Report 11/83, Den Haag, Concawe 1983.

[21] Heywood, J.B., "Future Engine Technology: Lessons From the 80's for the 1990's," ASME 1990 Soichiro Honda Lecture, ICE-Vol. 13, pp. 1-14, presented at the 12th Annual Fall Technical Conference of the ASME Internal Combustion Engine Division, Rockford, Illinois, October 7-10, 1990.

[22] *Automotive News*, "Mitsubishi: Weight, Not Size, is MPG Ticket," p. 1, April 1, 1991.

[23] *Automotive News*, "Bush's Dual-Fuel Plan Spells CAFE Relief," p. 2, February 18, 1991.

[24] *Automotive News*, "With War as Ammunition, Bryan Shoots for a CAFE Win," p. 3, February 4, 1991.

[25] *USA Today*, "Use Law to Demand More Efficient Cars," April 4, 1991.

[26] 40 CFR, Part 86, Subpart A, para. 86.113-82, p. 482, *Protection of the Environment*, July 1, 1985.

[27] Frank, M.E., "Gasoline Volatility Regulations: the Impact on the US Refining Industry," *Hydrocarbon Technology International*, 1990.

[28] Warner-Selph, M.A. and Smith, L.R., "Assessment of Unregulated Emissions from Gasoline Oxygenated Blends," EPA Report No 460/3-91-002, March 1991.

[29] *Oil & Gas Journal*, "New Transportation Fuels: The Challenges Ahead," OGJ Special, November 13, 1989.

[30] Lowi, A.,Jr. and Carter, W.P.L., "A Method for Evaluating the Atmospheric Ozone Impact of Actual Vehicle Emissions," SAE Paper 900710 presented at the SAE International Congress and Exposition, Detroit, Michigan, February 26 - March 2, 1990.

[31] Lenane, D.L., "Effect of a Fuel Additive on Emission Control Systems," SAE Paper 902097 presented at the SAE International Fuels and Lubricants Meeting and Exposition, Tulsa, Oklahoma, October 22-25, 1990.

[32] Clean Air Act Amendments of 1990 Senate Bill 1630 as passed by the U.S. House of Representatives and Joint House-Senate Conference Agreement, October 10, 1990, Title II, signed by President George Bush, November 5, 1990. Pages 77-145.

[33] Resolution 90-58 State of California Air Resources Board Agenda Item 90-14-1 Low Emission Vehicles/Clean Fuels Amending Title 13 California Code of Regulations, September 28, 1990.

[34] Heimrich, M.J., "Air Injection to an Electrically-Heated Catalyst for Reducing Cold-Start Benzene Emissions from Gasoline Vehicles," SAE Paper 902115 presented at the SAE International Fuels and Lubricants Meeting and Exposition, Tulsa, Oklahoma, October 22-25, 1991.

[35] Heimrich, M.J., Albu, S., and Osborn, J., "Electrically-Heated Catalyst System Conversions on Two Current-Technology Vehicles," SAE Paper 910612 presented at the SAE International Congress and Exposition, Detroit, Michigan, February 25-March 1, 1991.

[36] *Automotive News*, "GM Develops Group to Build and Market Electric Car Line," p. 1, March 11, 1991.

[37] 40 CFR, Part 50, Ch.1, 7-1-87 Edition, p.538-541.

[38] Colorado Air Quality Control Commission Regulation No. 13 implemented January 1, 1988.

[39] *Auto/Oil Air Quality Improvement Research Program*, Technical Bulletin 1, Initial Mass Exhaust Emission Results from Reformulated Gasolines, December 1990.

[40] Reuss, L.E., General Motors Corporation News Release Regarding Use of Reformulated Gasoline, March 18, 1991.

[41] Springer, K.J., "Diesel Lube Oils - Fourth Dimension of Diesel Particulate Control." ASME Paper presented at the 1988 Fall Technical Conference of ASME, Engine Emission Technology for the 1990s ICE Vol. 4, Book No. I00282, San Antonio, Texas, October 2-5, 1988. Also ASME Journal of Engineering for Gas Turbines and Power, Vol. 111, No. 3, pp. 355-360, July 1989.

[42] Springer, K.J., "Low Emission Diesel Fuel for 1991-1994." ASME Paper presented at the Internal Combustion Engine Symposium of the Twelfth Annual Energy-Sources Technology Conference & Exhibition, Advances in Engine Emissions Control Technology ICE Vol. 5, Book No. H00444-1989, January 22-25, 1989, Houston, Texas. Also ASME Journal of Engineering for Gas Turbines and Power, Vol. 111, No. 3, pp. 361-368, July 1989.

[43] Recommended Federal On-Highway Diesel Fuel Specifications to Assist Engine Manufacturers in Meeting the 1991 and 1994 Particulate Standards, submitted by American Petroleum Institute, National Petroleum Refiners Association, Engine Manufacturers Association, and National Council of Farmer Cooperatives, July 19, 1988.

[44] 40 CFR, Parts 80 and 86, Regulation of Fuels and Fuel Additives: Fuel Quality Regulations for Highway Diesel Fuel Sold in 1993 and Later Calendar Years: Final Rule, August 21, 1990.

[45] CRC Report No. 565, "A Program to Evaluate a Vehicle Test Method for Port Fuel Injector Deposit-Forming Tendencies of Unleaded Base Gasolines," February 1989.

[46] Richardson, C.B., Gyorog, D.A., and Beard, L.K., "A Laboratory Test for Fuel Injector Deposit Studies," SAE Paper 892116.

[47] Bitting, B., Gschwendtner, F., Kohlhepp, W., Kothe, M., Testroet, C.J., and Ziwica, K.H., "Intake Valve Deposits - Fuel Detergency Requirements Revisited," SAE Paper 872117 presented at the SAE International Fuels and Lubricants Meeting and Exposition, Toronto, Ontario, Canada, November 2-5, 1987.

[48] Springer, K.J., "Gasoline and Diesel Fuel Qualification - A National Need," ASME Paper 90-ICE-19 presented at the Thirteenth Annual Energy-Sources Technology Conference and Exhibition, New Orleans, Louisiana, January 14-18, 1990. Also ASME Journal of Engineering for Gas Turbines and Power, Vol. 112, pp. 398-407, July 1990.

[49] "Intake Valve Deposit/Driveability Investigations CRC IVD Group," Ford Engine, Fuels and Lubricants CAC; February 28, 1991.

[50] *Oil & Gas Journal*, "Additives to Have Key Role in New Gasoline Era," February 11, 1991.

[51] *Consumers Report*, "Which Gasoline for Your Car?" January 1990.

[52] Ullman, T., "Investigation of the Effects of Fuel Composition on Heavy-Duty Diesel Engine Emissions." SAE Paper 892072 presented at the SAE Fuels and Lubricants Meeting, Baltimore, Maryland, September 25-28, 1989.

[53] Springer, K.J., "Diesel Fuels and the Future of Diesels in the U.S.," presented at the Xth AGELFI European Automotive Symposium, Ostend, Belgium, October 11-12, 1990.

[54] Title 13, California Code of Regulations Section 2256, Aromatic Content of Diesel Fuel Subsection(g) Certified Diesel Fuel Formulations Resulting in Equivalent Emission Reductions, April 17, 1989, amended December 13, 1990, February 8 and April 15, 1991.

[55] Ullman, T.L., Mason, R.L., and Montalvo, D.A., "Effects of Fuel Aromatics, Cetane Number, and Cetane Improver on Emissions from a 1991 Prototype Heavy-Duty Diesel Engine," SAE Paper 902171 presented at the SAE International Fuels and Lubricants Meeting and Exposition, Tulsa, Oklahoma, October 22-25, 1991.

[56] Newkirk, M.S., Smith, L.R., and Ahuja, M., "Formaldehyde Emission Control Technology for Methanol-Fueled Vehicles," SAE Paper 902118 presented at the SAE Fuels and Lubricants Meeting and Exposition, Tulsa, Oklahoma, October 22-25, 1991.

[57] Springer, K.J., "Particulate Trap for Two-Stroke Cycle Detroit Diesel Powered City Bus," ASME Paper presented at the 1989 Fall Technical Conference of ASME, Engine Design, Operation and Control Using Computer Systems ICE Vol. 9, Book No. I00295-1989, Dearborn, Michigan, October 15-18, 1989.

ICE-Vol. 15, Fuels, Controls, and Aftertreatment
For Low Emissions Engines
ASME 1991

EFFECTS OF OXYGEN ENRICHMENT AND FUEL EMULSIFICATION ON DIESEL ENGINE PERFORMANCE AND EMISSIONS

R. R. Sekar, W. W. Marr, R. L. Cole, and T. J. Marciniak
Argonne National Laboratory
Argonne, Illinois

J. E. Schaus
Auto Research Laboratories, Incorporated
Chicago, Illinois

ABSTRACT

Argonne National Laboratory (ANL), in cooperation with AutoResearch Laboratories, Inc. (ALI), has completed a series of tests on a single-cylinder, direct-injection diesel engine coupled to an oxygen-enriching membrane system. The data from the first series of tests using bottled oxygen have been previously reported. That series of tests included no examination of the effects of changing the injection timing, which is an important operating parameter that affects NO_x emissions and efficiency. In the second test series, the subject of this paper, the effects of injection timing were investigated. In addition, an oxygen-enriching membrane was used to supply combustion air. Use of bottled oxygen in any real diesel engine application would be a safety hazard, so bottled oxygen is unlikely to be used in commercial engine applications. For this reason, it is important to demonstrate an engine system with an on-line oxygen-enriching device.

Tests were conducted with #2 and #4 diesel fuels. The data indicated that NO_x emissions increase when the oxygen level is increased from 21 to 27%, but retarding the injection timing by 11 degrees crank angle significantly reduced the NO_x emissions. The effect on NO_x reduction of retarding the injection timing is greater at higher oxygen levels. The water emulsification of the fuels also reduced NO_x emissions significantly (Sekar et al., 1990a).

It was shown in both series of tests that oxygen-enriched combustion air reduced particulate emissions, smoke, and ignition delay. The effect on ignition delay resulted in a favorable NO_x vs. fuel consumption trade-off when the injection timing was changed. The collective data lead to the conclusion that an optimum set of the major operating variables, including (1) oxygen level in the combustion air, (2) water level in the fuel, and (3) injection timing, could lead to a diesel engine system that has (i) significantly lower particulates, smoke, and NO_x emissions, without loss of efficiency; (ii) the ability to use lower-cost heavy liquid fuels; and (iii) the potential for increasing the power output by as much as 50% with only a nominal increase (about 15%) in peak cylinder pressure.

INTRODUCTION

The concept of using oxygen-enriched air for diesel engine combustion has been studied by several researchers over the last two decades. The main motivation for oxygen enrichment is to lower the smoke and other exhaust emissions, as well as to improve the thermal efficiency. Wartinbee (1971) considered the concept of oxygen enrichment for spark-ignition engines but rejected it due to the difficulties involved in controlling NO_x emissions caused by oxygen enrichment. Quader (1978) studied the combustion mechanisms of oxygen enrichment; the concept was again rejected due to the NO_x and the fuel-consumption penalties that were encountered. Ghojel et al. (1983) and Iida et al. (1986) published their work on indirectly injected and directly injected diesel engines, respectively. Later, Iida and Sato (1988) found that the increased NO_x could be controlled by retarding the injection timing, which is made possible by the reduced ignition delay.

The application of the oxygen-enrichment concept to diesel engines is being reviewed in the context of particulate emissions standards proposed by the U.S. Environmental Protection Agency (EPA) that will go into effect in 1994. Since oxygen enrichment can potentially reduce smoke and particulate emissions significantly, this technology may provide a new option to solve diesel engine emissions problems. Recent work by engine developers (Watson et al., 1990; Willumeit and Bauer, 1988) indicates a renewed interest in the concept.

In parallel developments, funded primarily by the U.S. Department of Energy (DOE), significant advances were reported in practical oxygen-enrichment devices, such as "asymmetric hollow fiber" membranes that could be used for various end-use applications (Whipple and Ragland, 1989; Gollan and Kleper, 1985; Kobayashi, 1987). Argonne National Laboratory (ANL) undertook a systematic research project on the application of oxygen enrichment to stationary diesel engines. Although the concept offers several advantages in terms of performance and emissions, oxygen-enriched diesel engines cannot be commercialized in stationary or transportation applications in the U.S. without a definite means of controlling the NO_x emissions. Hence, the use of water injection, in the form of emulsified fuel, was included as part of the ANL research. Water injection has been previously studied and reported to reduce NO_x (Valdmanis and Wulfhorst, 1970; Greeves et al., 1976). Analytical studies (Assanis et al., 1990; Cole et al., 1990) of the performance, emissions, and economic aspects of the diesel engine with oxygen enrichment and emulsified fuels have revealed that significant decreases in smoke, particulates, and other emissions (except NO_x), as well as decreased ignition delays, are possible. At the same time, excellent increases in power density and slight increases in efficiency could be achieved. In the next phase of the project, ANL conducted tests on a single-cylinder diesel engine to obtain performance, emissions, and cylinder-pressure data. Data from the first series of engine tests, in which bottled oxygen was used to increase the oxygen level of the combustion air, have been previously published by ANL (Sekar et al., 1990a; Sekar et al., 1990b; Sekar et al., 1990c). Although that test series was very useful for quantifying the benefits and the challenges to engine performance and emissions, two major items could not be included. The effects of fuel injection timing and use of an actual on-line oxygen-enriching membrane as part of the engine system should be investigated. This report documents these two aspects of the research.

OBJECTIVES

The main objective of the series of tests considered in this paper was to show that an oxygen-enriching membrane can operate as part of a diesel engine system. It was intended to repeat some data points from the first series of tests by Sekar et al. (1990a, b, and c) to confirm the benefits of and problems with oxygen-enrichment, fuel emulsification, and use of heavier base fuels. The second objective was to obtain performance, emissions, and cylinder-pressure data at three different fuel injection timings to determine the effect of fuel injection timing on engine performance and on the combustion process.

EXPERIMENTAL SET-UP

Engine and Fuels Used

A single-cylinder, four-stroke, direct-injection diesel engine was used in this series of experiments. This is a one-cylinder version of a heavy-duty diesel engine commonly used in on-highway trucks and other applications. The major specifications of the base engine are given in Table 1. No hardware changes were made to the base engine, and the manufacturer's recommendations were used in the set-up and operating procedures.

The engine is designed to run on #2 diesel fuel. The objective of this project is to test the engine on a less refined fuel, such as #6 diesel. Because it was felt that the engine might not run smoothly on # 6 diesel fuel without extensive modifications, it was decided to compromise and test the engine on # 4 fuel, which is generally used in marine applications. Introduction of water into the combustion process was accomplished by emulsifying the two base fuels with distilled water and a small percentage of stabilizing chemical additive. Three levels of water content were tested with each base fuel.

The Oxygen Supply System

An oxygen-enriching membrane was available from a previous project (see Whipple and Ragland, 1989; Gollan and Kleper, 1985).

Table 1 Test Engine Specifications

Parameter	Value
Number of Cylinders	1
Bore (mm) × Stroke (mm)	137 × 165
Displacement (L)	2.44
Engine Speed (rpm)	1800
Injection Timing (°btdc)[1]	33
Compression Ratio	14.5
Peak Cylinder Pressure (bar)	110

A schematic diagram of the air supply system is shown in Fig. 1. An air compressor was used with a cooler and dehumidifier to supply feed air to the membrane. The nitrogen-rich air was discarded, and the oxygen-rich air and the shop air were mixed in a large tank before the intake manifold of the engine. A micro-fuel-cell type (Teledyne model 326A) oxygen sensor located in the engine intake manifold was used to measure the intake oxygen content of the air entering the engine. Elaborate safety systems were provided to handle oxygen. The engine crank case was purged with nitrogen for added safety while running the engine. All the other test set-up and measurement details were exactly the same as in the first series of tests (Sekar et al., 1990a, b, and c).

Test Matrix

Two engine operating conditions were tested in this series. "50% Load" is defined as the engine operating conditions (intake manifold pressure, exhaust manifold pressure, and mass flow rate of air plus oxygen) corresponding to 50% brake power level of the base engine, which is 18.65 kW (25 hp). "100% Load" is defined as the engine operating conditions corresponding to the rated power level of the base engine, which is 37.3 kW (50 hp). The intake manifold pressures were maintained at 112 cm Hg. abs at the "50% load" point and 140 cm Hg abs. at the "100% load" point. The exhaust manifold pressure was maintained at 81 cm Hg abs. throughout. Only constant power output data were obtained this time, since the cases of constant exhaust-to-intake-oxygen ratio and constant exhaust-oxygen level already have been reported (Sekar et al., 1990a, b, and c). Only one intake-oxygen level, besides the base of 21%, was used. The membrane has the selectivity to produce up to 35% oxygen, but the permeability was not adequate to provide the mass flow rate required by the engine to produce constant power. Hence, shop air was mixed with oxygen-enriched air from the membrane unit. The power had to be constant, so the intake-oxygen level of this system was adjusted to establish the equilibrium condition; in most test runs, this was at 25-27% oxygen in the intake air. Fuel emulsions were used with water contents of 0 and 5% by weight. Seventy-two test runs were made, and many graphs were plotted. Only a portion of the curves, showing most significant trends, are presented in this report. Overall, the test series was completed without any major engine problems.

[1]btdc = before top dead center. A timing of N degrees crank angle before top dead center is commonly indicated by N° btdc.

The base injection timing was 33° btdc. Two retarded timings of 27° and 22° btdc were used in this test matrix to investigate the effects of varying injection timing on engine performance and emissions.

DISCUSSION OF THE DATA

Membrane Performance

Initially the membrane alone was tested, with the engine shut off, to map the flow rate (permeability) and oxygen purity (selectivity) that could be achieved with the available membrane. (This membrane was not optimized for use with a diesel engine or for oxygen-enriched air as the product. The membrane was originally purchased by DOE for a membrane-development project in which the product was pure nitrogen and the oxygen-enriched air was just a waste stream.) Figure 2 illustrates the membrane performance. The membrane is capable of producing up to 35% oxygen, but the flow rate of oxygen-enriched air is low (only about 30% of the inlet air flow rate). This figure shows two major problems connected with the application of membrane technology to engines. First, the high pressure levels used in the experiment are generally not available in an engine application. Second, the "yield" of oxygen-enriched air, compared to the membrane inlet air flow rate, needs to be improved. Such an improvement would reduce both the parasitic losses and the cost.

Engine Performance

The major independent variable of interest in this part of the investigation was injection timing; all the data are presented as a function of injection timing. As shown in Fig. 3, the thermal efficiency decreases as the injection timing is retarded from 33° to 22° btdc. This decrease is a well-known characteristic of diesel engines. However, the same data at a higher oxygen level seem to indicate a different optimal timing, at approximately 27°. The thermal efficiency is slightly, but consistently, higher at higher oxygen levels. Figure 4 illustrates a similar trend when the fuel was emulsified with 5% water. The gains in power density potential are expected to be the same as those reported earlier (Sekar et al., 1990a, b, and c). It is important to increase the engine power rating when oxygen-enriched air is used. Part of the increased power is needed to offset the parasitic losses of the membrane. At the 50%-load point and with a #4 base fuel, similar trends in performance were noted.

Emissions

One of the conventional means of controlling NO_x emissions is to retard the injection timing. However, when the timing is retarded the brake specific fuel consumption (bsfc) increases. Hence, the actual engine timing is determined as a trade-off between NO_x and bsfc. The effect of injection timing on NO_x emisions is presented in Fig. 5. The reduction in NO_x observed in the base case (21% oxygen), as well as in the higher-oxygen-level case (25% oxygen), followed predictable trends. At 25% oxygen the effect of injection timing on NO_x emissions was slightly more pronounced than in the base case. Figure 6 is a typical NO_x vs. bsfc trade-off curve. At higher oxygen levels, retarding the injection timing has a large impact on NO_x but very little impact on bsfc. This could be explained by the reduction in ignition delay when oxygen-enriched air is used for combustion. This is a characteristic that could be used to advantage when the engine is optimized for oxygen-enriched combustion air. Similarly, NO_x emissions and particulate emissions are also controlled in a trade-off mode. When injection timing is retarded to decrease NO_x emissions, particulate emissions increase. However, as shown in Fig. 7, when oxygen-enriched air is used, the effect on particulate emissions of retarding the timing is insignificant compared to the effect with standard air. The results regarding the use of neat #4 diesel fuel and fuel-water emulsions confirmed the earlier trends.

Combustion

Combustion in diesel engines is usually analyzed by measuring the cylinder pressure as a function of crank angle. Figure 8 shows the effect of injection timing on cylinder pressures during the combustion portion of the cycle. Figure 9 presents the corresponding data at a higher oxygen level. Comparison of these two figures suggests that an oxygen-enriched diesel engine is likely to behave exactly the same as a conventional diesel engine. Corresponding heat release diagrams are presented in Figs. 10 and 11. Again, these figures indicate very close similarities between a conventional and an oxygen-enriched diesel engine. In terms of technology transfer, this is encouraging, because it implies that engine modifications will be minor; thus, engine manufacturers are more likely to consider these technologies for incorporation into their products. The effects of oxygen enrichment on peak cylinder pressure and the smoke-limited gross engine power output are shown in Fig. 12. These data were obtained in the first series of tests using bottled oxygen. Given a nominal increase in cylinder pressure, one obtains a dramatic increase in power density potential. This indicates that engine manufacturers might be able to incorporate oxygen-enrichment technology without major engine changes.

CONCLUSIONS

1. Oxygen enrichment of combustion air is a viable technology for diesel engines. The demonstration of a membrane separator to enrich the oxygen as part of a diesel engine system proves that the concept is technically feasible. No major engine modification is needed to incorporate the membrane system.

2. The remarkable increase in power density potential demonstrated in these tests is important in industrial cogeneration and power generation plants. The oxygen enrichment system has considerable parasitic losses, and part of the increased power from the engine will be utilized to overcome these losses. The net power gain should help to reduce the capital cost of the plant and to offset the capital cost of the membrane system.

3. Oxygen enrichment slightly improves thermal efficiency, and hence fuel consumption.

4. Oxygen enrichment, by itself, increases NO_x emissions if the power output of the engine remains unchanged. Retarding the injection timing and introducing water into the combustion process can significantly reduce the NO_x emissions.

5. Particulate emissions data are mixed. The base engine has a low level of mass particulate emissions. The measurement accuracy of changes in particulate emissions due to oxygen enrichment was not adequate to give consistent data. The general trend is for the particulate matter to decrease with increasing oxygen levels in the air.

6. Oxygen enrichment and the resulting higher power output increases the cylinder pressures. The peak pressure is increased by about 15% when the power is increased by 140%.

7. Both oxygen enrichment and water emulsification of fuel show a definite trend in the heat release diagrams. The higher proportion of the energy release occurs in the early part of the combustion process. This trend should be taken into consideration when the combustion process geometry and parameters are optimized.

8. No attempt was made in these experiments to optimize the operating variables. By employing these two technologies, i.e., oxygen enrichment and fuel emulsification, an engine designer will have two more degrees of freedom to improve the overall system performance and emissions.

ACKNOWLEDGMENT

The work reported in this paper was supported by the U.S. Department of Energy, Assistant Secretary for Conservation and Renewable Energy, Office of Industrial Technologies, under contract W-31-109-Eng-38.

REFERENCES

Assanis, D.N., Sekar, R.R., Baker, D., Siambekos, C.T., Cole, R.L., and Marciniak, T.J., 1990, "Simulation Studies of Diesel Engine Performance with Oxygen Enriched Air and Water Emulsified Fuels," American Society of Mechanical Engineers Paper No. 90-ICE-17, published at the Energy-Sources Technology Conference and Exhibition, New Orleans, LA (Jan.).

Cole, R.L., Sekar, R.R., Stodolsky, F.S., and Marciniak, T.J., 1990, "Technical and Economic Evaluation of Diesel Engine with Oxygen Enrichment and Water Injection," American Society of Mechanical Engineers Paper No. 90-ICE-1 (Jan.).

Ghojel, J., Hillard, J.C., and Levendis, J.A., 1983, "Effect of Oxygen-Enrichment on the Performance and Emissions of I.D.I. Diesel Engines," Society of Automotive Engineers Paper No. 830245, Detroit, MI.

Gollan, A.Z., and Kleper, M.H., 1985, "Research into an Asymmetric Membrane Hollow Fiber Device for Oxygen-Enriched Air Production (Phase II Final Report)," U.S Department of Energy Report No. DOE/ID/12429-1.

Greeves, G., Khan, I.M., and Onion, G., 1976, "Effects of Water Introduction on Diesel Engine Combustion and Emissions," *Sixteenth Symp. (International) on Combustion*, The Combustion Institute, Pittsburgh, PA, pp. 321-336 (Aug.).

Iida, N., et al., 1986, "Effects of Intake Oxygen Concentration on the Characteristics of Particulate Emissions from a D.I. Diesel Engine," Society of Automotive Engineers Paper No. 861233, Detroit, MI.

Iida, N., and Sato, G.T., 1988, "Temperature and Mixing Effects on NO_X and Particulate," Society of Automotive Engineers Paper No. 880424, Detroit, MI.

Kobayashi, H., 1987, "Oxygen-Enriched Combustion System Performance Study," U.S. Department of Energy Report No. DOE/ID/12597.

Quader, A.A., 1978, "Exhaust Emissions and Performance of a Spark-Ignition Engine Using Oxygen-Enriched Intake Air," *Combustion Science and Technology*, Vol. 19, pp. 81-86.

Sekar, R.R., Marr, W.W., Cole, R.L., Marciniak, T.J., and Schaus, J.E., 1990a, "Diesel Engine Experiments with Oxygen-Enrichment, Water Addition and Lower-Grade Fuel," presented at 25th Intersociety Energy Conversion Engineering Conference, Reno, NV (Aug.)

Sekar, R.R., Marr, W.W., Cole, R.L., Marciniak, T.J., and Schaus, J.E., 1990b, "Cylinder Pressure Analysis of a Diesel Engine Using Oxygen-Enriched Air and Emulsified Fuels," presented at Society of Automotive Engineers (SAE) International Off-Highway and Powerplant Congress and Exposition, Milwaukee, WI (Sept.), SAE Paper No. 901565.

Sekar, R.R., Marr, W.W., Assanis, D.N., Cole, R.L., Marciniak, T.J., and Schaus, J.E., 1990c, "Oxygen-Enriched Diesel Engine Performance: A Comparison of Analytical and Experimental Results," presented at the 12th Annual Fall Technical Conference of the American Society of Mechanical Engineers (ASME), Internal Combustion Engine Division, Rockford, IL (Oct.), ICE-Vol. 13.

Valdmanis, E., and Wulfhorst, D.E., 1970, "The Effects of Emulsified Fuels and Water Induction on Diesel Combustion," Society of Automotive Engineers Paper No. 700736, Detroit, MI.

Wartinbee, W.J., Jr., 1971, "Emissions Study of Oxygen-Enriched Air," Society of Automotive Engineers Paper No. 710606, Detroit, MI.

Watson, H.C., Milkins, E.E., and Rigby, G.R., 1990, "A New Look at Oxygen Enrichment (1) The Diesel Engine," Society of Automotive Engineers Paper No. 900344, Detroit, MI.

Whipple, J.G., and Ragland, K.W., 1989, "Testing and Evaluation of Hollow Polymeric Fiber Membrane Air Separation Systems," U.S. Department of Energy Report (April).

Willumeit, H.-P., and Bauer, M., 1988, "Emissions and Performance of an S.I. Engine Inducting Oxygen-Enriched Combustion Air" (in German), *MTZ Motortechnische Zeitschrift,* Vol. 49, No. 4, pp. 149-152.

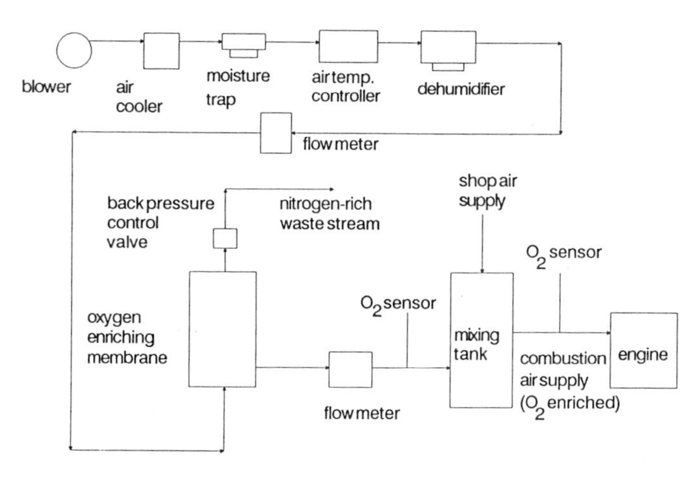

FIG. 1 Schematic Diagram of Oxygen Supply System for Engine Tests

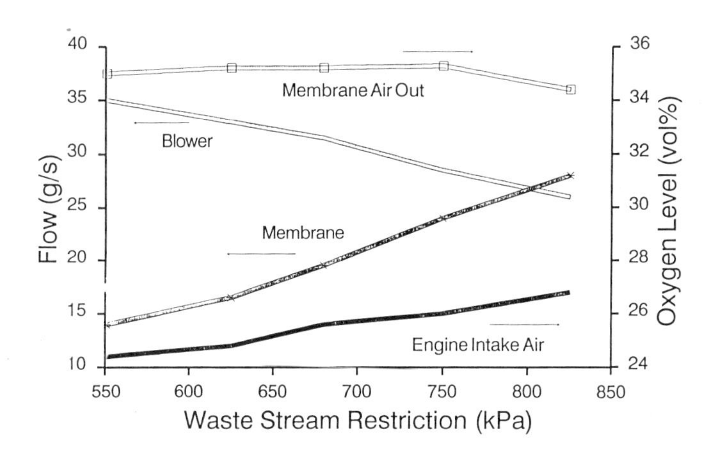

FIG. 2 Performance Map for the Air Separation Membrane

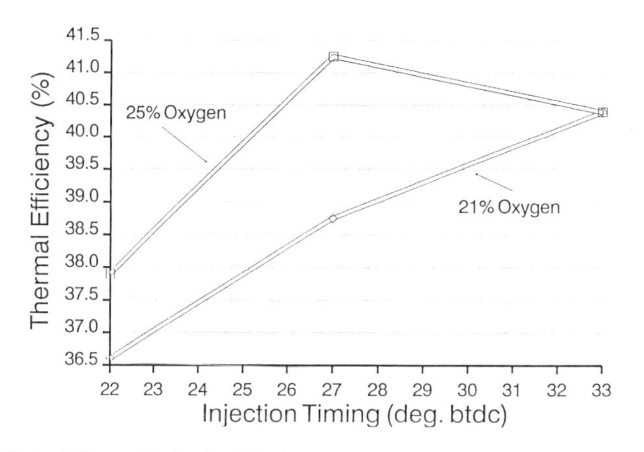

FIG. 3 Effect of Injection Timing on Thermal Efficiency -- Oxygen-Enriched Air and Standard Fuel, 50 bhp, No Water

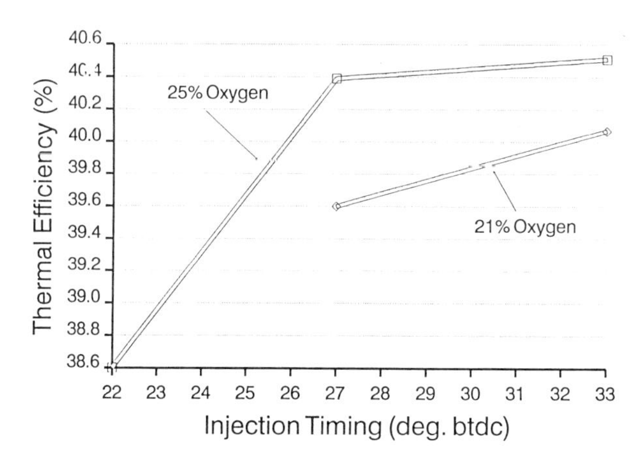

FIG. 4 Effect of Injection Timing on Thermal Efficiency -- Oxygen-Enriched Air and No. 2 Diesel Fuel Emulsified with 5% Water, 50 bhp

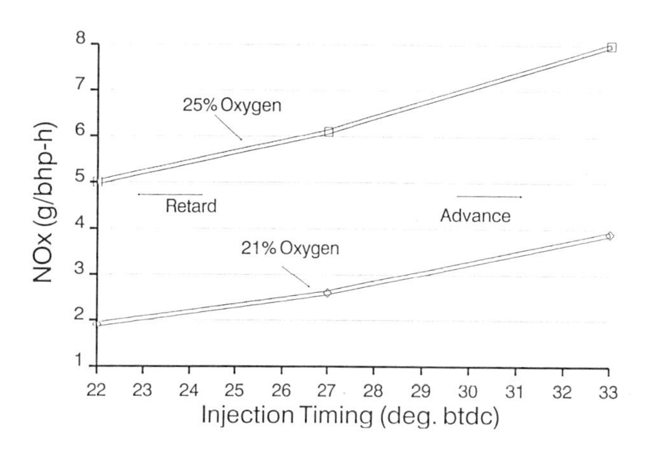

FIG. 5 Effect of Injection Timing on NO_x, 50 bhp, No. 2 Diesel Fuel, No Water

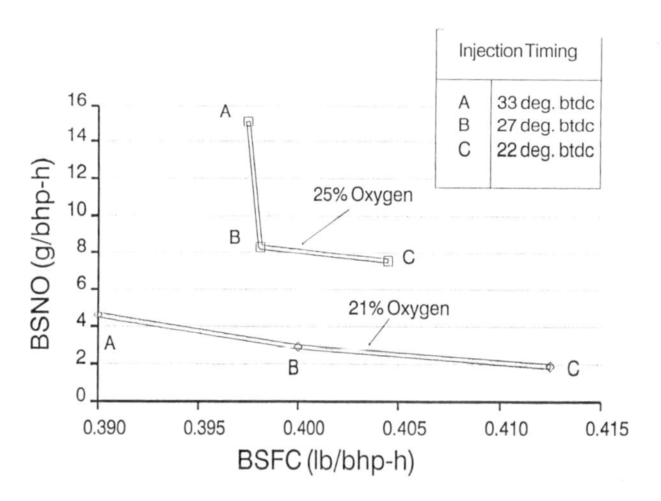

FIG. 6 Trade-Off between Brake Specific Fuel Consumption (bsfc) and NO_x ($BSNO_x$) -- No. 2 Diesel Fuel, 25 bhp, No Water

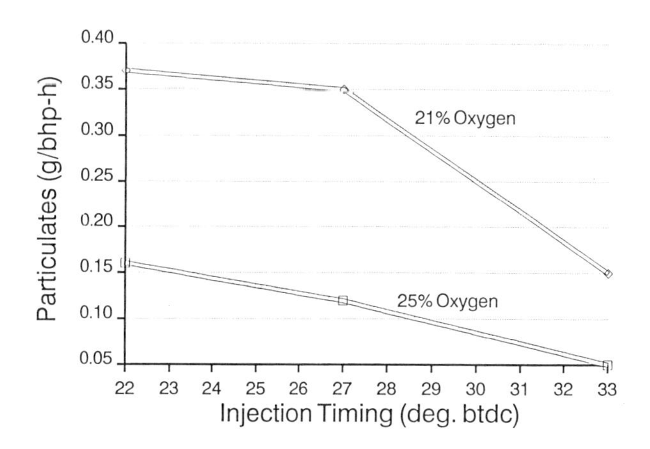

FIG. 7 Effect of Injection Timing on Particulate Emissions -- Oxygen-Enriched Air and No. 2 Diesel Fuel, 50 bhp, No Water

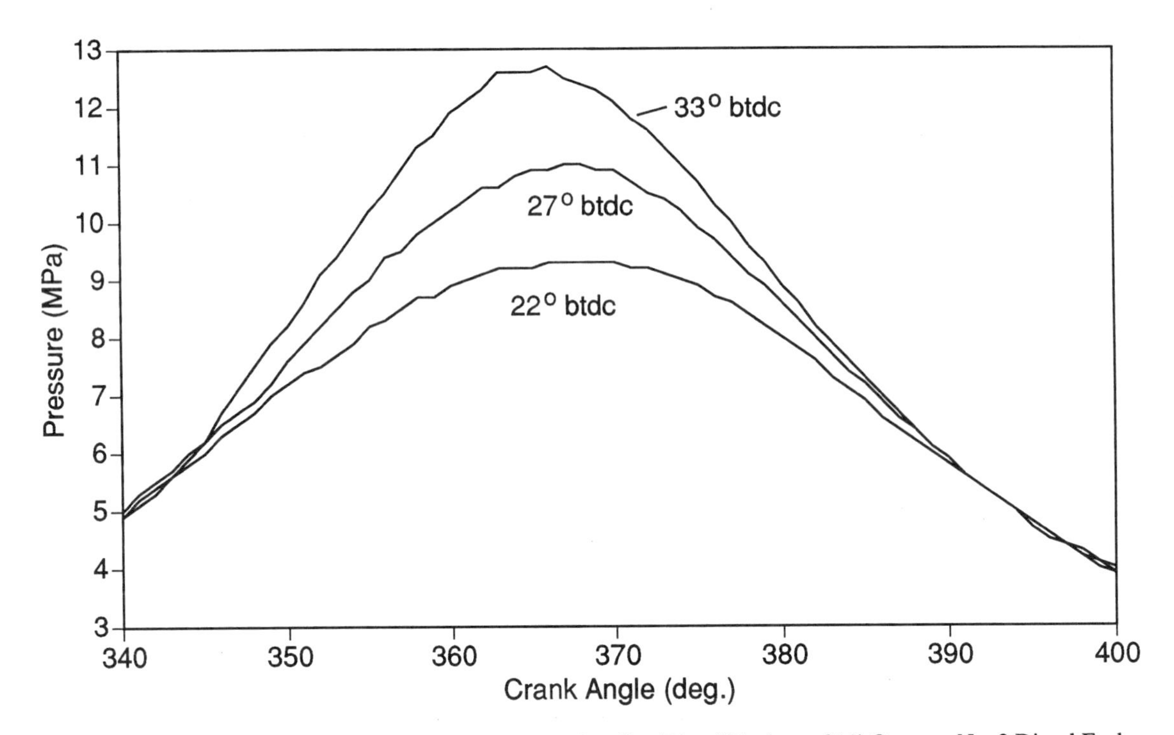

FIG. 8 Effect of Injection Timing on Cylinder Pressure, Baseline Diesel Engine -- 21% Oxygen, No. 2 Diesel Fuel, Full Load, No Water

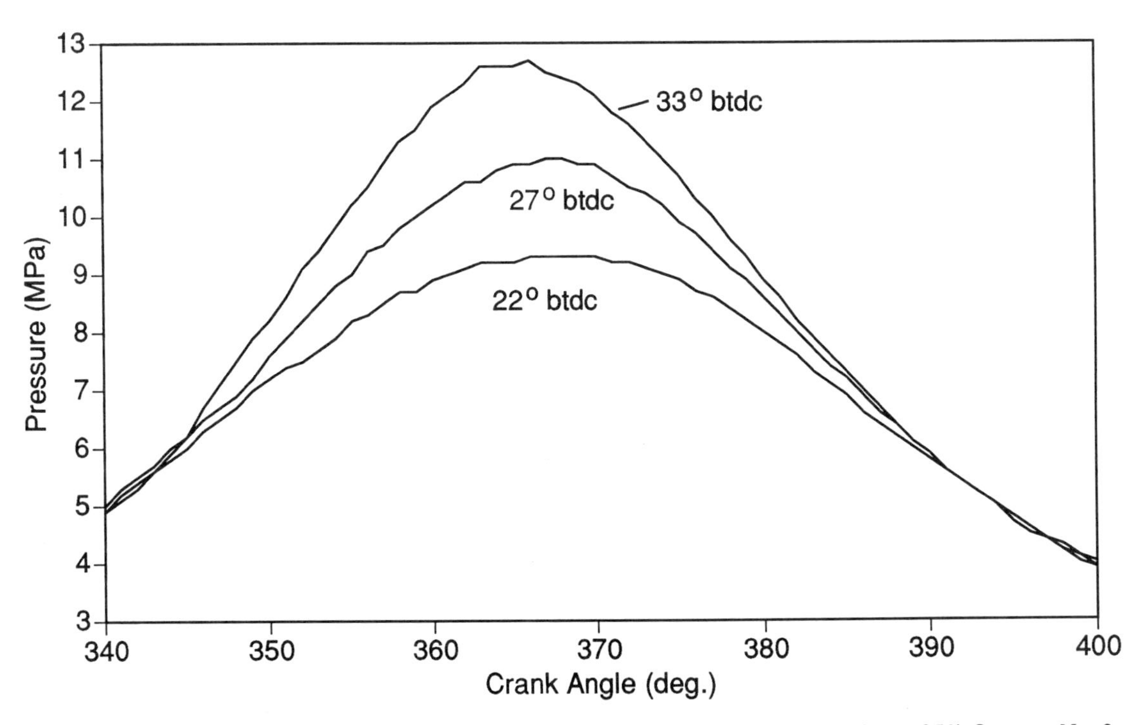

FIG. 9 Effect of Injection Timing on Cylinder Pressure, Oxygen-Enriched Diesel Engine -- 25% Oxygen, No. 2 Diesel Fuel, Full Load, No Water

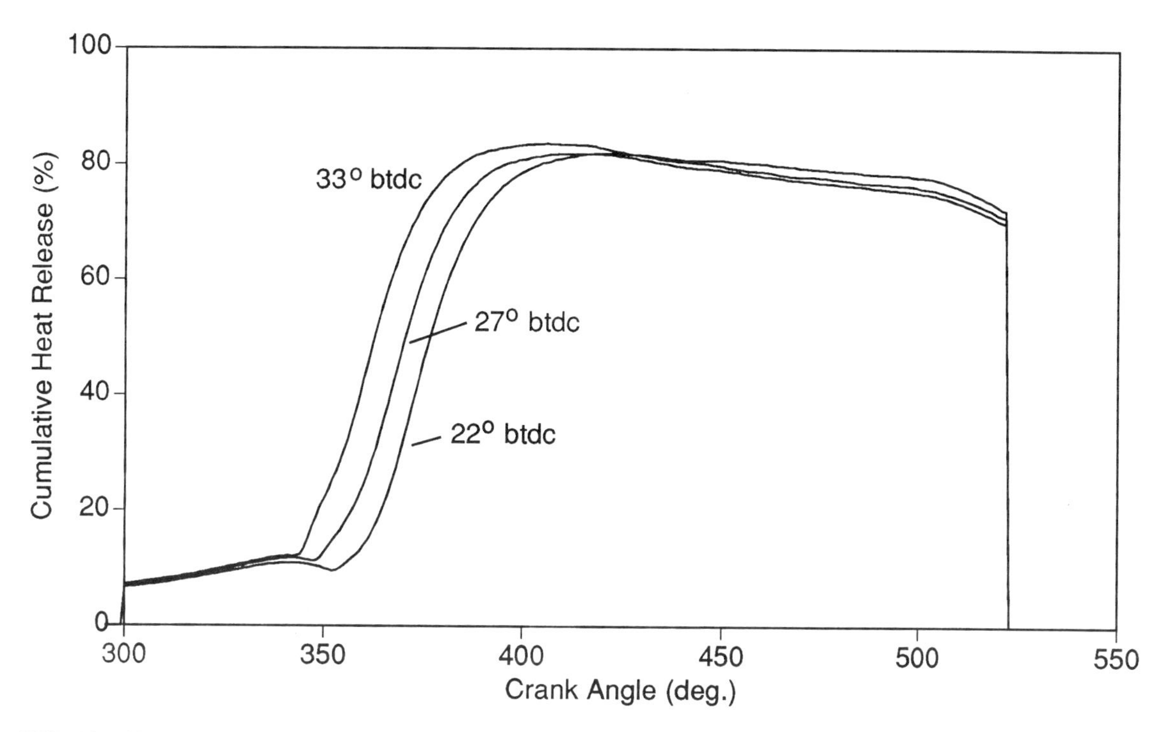

FIG. 10 Effect of Injection Timing on Cumulative Heat Release Rate, Baseline Diesel Engine -- 21% Oxygen, No. 2 Diesel Fuel, Full Load, No Water

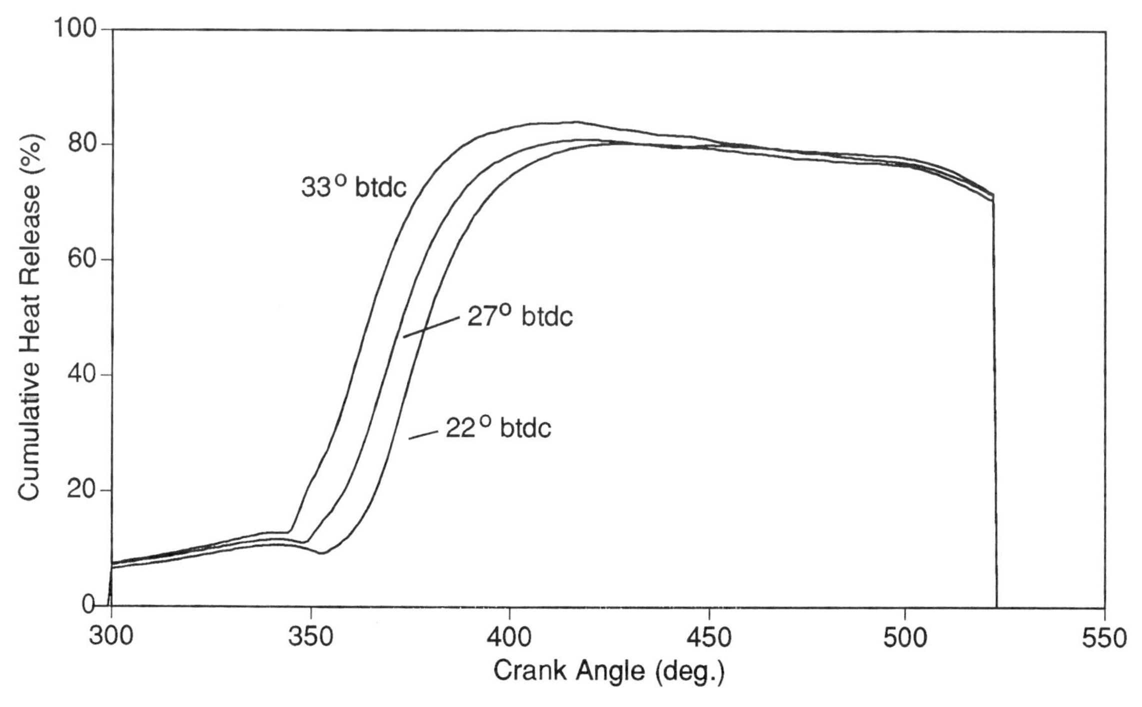

FIG. 11 Effect of Injection Timing on Cumulative Heat Release Rate, Oxygen-Enriched Diesel Engine -- 25% Oxygen, No. 2 Diesel Fuel, Full Load, No Water

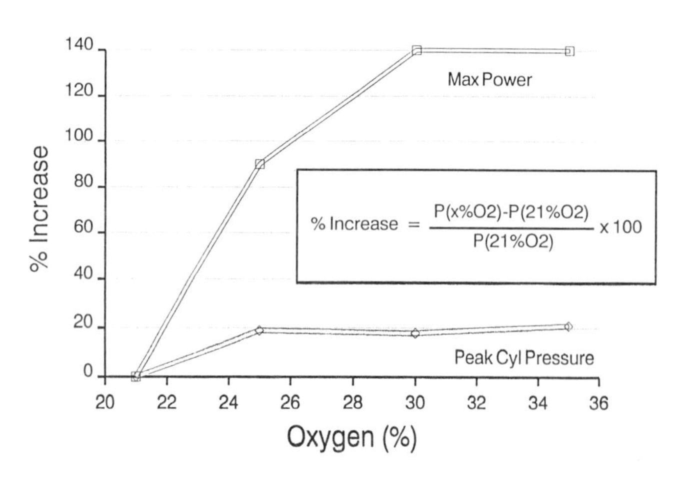

FIG. 12 Effect of Oxygen Level on Maximum Power Output and Peak Cylinder Pressure

ICE-Vol. 15, Fuels, Controls, and Aftertreatment
For Low Emissions Engines
ASME 1991

GAS ENGINES OR FUEL CELLS FOR FUTURE LOW EMISSIONS COGENERATION?

J. R. Thomas
Ricardo Consulting Engineers Limited
Shoreham-by-Sea, United Kingdom

M. R. Fry
W. S. Atkins Energy
Epsom, United Kingdom

ABSTRACT

In the light of continuing demands for energy conservation and pollution control, the paper reviews Gas Engines and Fuel Cells, examining their potential. The focus is the power range 50-3000 kWe. Combustion technologies are discussed for both spark and diesel pilot ignited engines and the importance of aftertreatment is seen. The need for better gas engine combustion modelling is stated. Continued progress is expected.

The various types of Fuel Cell are described and their merits, costs and potential are discussed in the context of a developing market. The effort now being invested indicates that Fuel Cells could provide viable, efficient sources of electricity and heat early next century. A price of $1000/kWe (1989 values) is seen as the target for market entry.

1. INTRODUCTION

Cogeneration has become both economically attractive and environmentally favourable in the past decade. In cases where recovered heat can be used the environmental dividend is very real and so a market has developed for relatively small power plants in the 50-3000 kWe range. To meet the strict objectives for air quality standards, Natural Gas has offered important advantages which, combined with its abundance and reasonable price, have made it the best fuel for cogeneration, both in and above this size range.

The development of the market for cogeneration has coincided with stricter regulation of exhaust emissions. In most countries, including the USA, the emphasis has been on reduced NOx emissions. The spark-ignited natural gas engine has proved to be the most successful prime mover for cogeneration in the power range mentioned above, using lean-burn or catalyst technologies to achieve the required low NOx performance with acceptable brake thermal efficiency. The dual fuel engine too has had a small share in this market. Better gas/air mixing, higher turbocharging efficiency, high precision high energy ignition systems, better charge cooling and improved combustion system designs have all contributed to the progress made. Major limitations are still detonation and, for the SI engines, the achievement of reliable, repeatable ignition with high durability.

Can the reciprocating gas engine be further improved? If so, by how much? Are major changes foreseen, with new powerplants replacing reciprocating engines further and further down the power range? This paper examines the future of the reciprocating gas engine in the context of a future challenger, the Fuel Cell. For cogeneration applications this is seen by some as the total answer to the needs of the first half of the 21st Century.

2. GAS ENGINES FOR FUTURE COGENERATION

The main issues affecting the future of gas engines for generation are the technical ones of meeting more stringent environmental regulations and continuing to improve shaft efficiency without excessive degradation of the recoverable exhaust heat. The investment required to meet these objectives is not insignificant but there is already great impetus in this direction. In the case of the low emissions natural gas engine, the necessary technology is not an automatic spin-off from automotive work and new technical initiatives will also be required. The cost of future engines in this field will undoubtedly increase as they become more sophisticated. If Fuel Cells are to challenge the gas engine it is thought by their developers that their cost must fall to $1000/kWe at 1989 values. The cost of small CHP packages does reach this figure today but units above about 150 kWe cost much less. It is expected that more rationalisation and an expanding market will enable package prices to fall by the year 2000, despite an increase in engine prices.

2.1 Spark-Ignited Engines

The limitations imposed by detonation and combustion stability have already been mentioned. For premixed combustion the potential for increasing output by turbocharging and for increasing brake thermal efficiency by increasing the compression ratio is severely limited. NOx reducing measures such as charge cooling, dilution with air or EGR, timing retard and compression ratio reduction also serve to reduce the knocking potential, but the trade-offs with efficiency and some other emissions are not favourable. Combustion chamber and ignition system optimisation enable performance improvements to be achieved, but the evolution of these improvements will become slower and slower. Engine management systems will enable engines to be run closer to the margins but further improvements are required in λ-sensing and the discernment of potentially damaging detonation.

The lean-burn approach offers excellent low NOx performance but further improvements are strongly dependent upon the development of better ignition systems and the management of gas/air mixing. If Total Hydrocarbon (THC) emissions become subject to regulations, there must be progress in breaking down the strong correlation between excess air ratio and THC emissions (Figure 1). It is important to note that very lean engines can be run with excellent cycle-to-cycle stability and good efficiency and still emit 2500 ppm of Methane. In this context, the development of Catalysts with better Methane conversion at lean-burn engine exhaust temperatures must proceed.

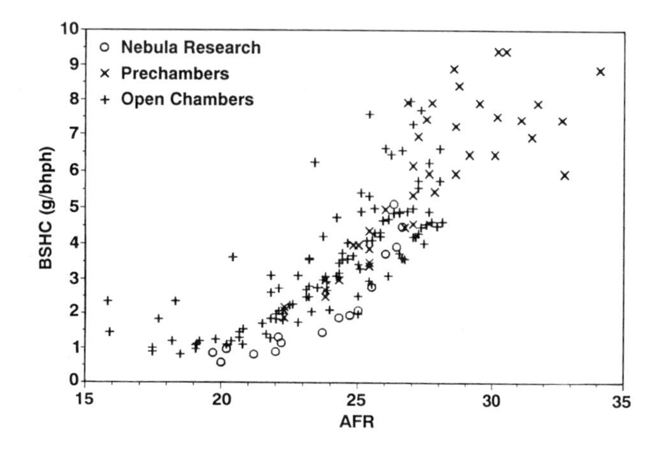

Figure 1 - Effect of Air/Fuel Ratio (AFR) on Total HC for Lean-Burn Engines

The concept of the small rich prechamber is considered an important one for the continued development of the diluted charge engine. The prechamber serves as a "spark plug amplifier" to help overcome the difficulties of igniting lean mixtures and then achieving rapid flame propagation. Although ignition systems able to achieve higher voltages and larger energies have been developed in recent years, these cannot provide for mixtures at $\lambda > 1.5$ with little turbulence. (Indeed it is found that they are only just adequate to ignite the rich prechamber mixtures when the boost pressure and/or compression ratio is high). The energy emitted from prechambers is many orders of magnitude greater and fast combustion of lean mixtures up to $\lambda \geq 2$ can be achieved. The NOx which results from both the combustion within the prechamber and that of fuel-rich material emitted from the prechamber ahead of the flame, becomes significant when very low emissions are being sought. Current emphasis must therefore be on reducing the size of the prechamber and controlling its mixing and combustion more effectively. It is becoming important to identify the lowest limit of prechamber volume (or fuel charge) for a given engine cylinder size.

With good management of prechamber charging and mixing, volumes well below 1% of the clearance volume should be viable for engines above 200 mm bore. Other forms of "spark plug amplifier" may also emerge to provide high ignition energy for lean charges and with further improvements and control of mainchamber mixture quality it is expected that lean-burn gas engines will achieve in-service NOx emissions of 0.5 g/kWh without catalysts. Without some major increase in available ignition energy it is not thought that diluted charge open chamber engines will improve upon their current best NOx/THC/Efficiency levels by more than a few percent in the coming years, except with aftertreatment.

The 3-way catalyst process is expected to remain as an important means of simultaneously reducing NOx, CO and HC emissions to the lowest levels. The need for improved Methane conversion is being addressed and in the shaft power range below 500-1000 kW the stoichiometric engine is expected to remain viable despite the bmep limitation imposed by thermal load and knock considerations. The use of EGR as a diluent gas can enable these barriers to be moved back somewhat, whilst still retaining the 3-way catalyst capabilities [1]. The specific heat capacity of exhaust gas with its high CO_2 content makes it a more effective diluent than air if it can be adequately cooled [2] but the requirement for good mixing would now demand that three components be intimately mixed before ignition occurs. More investigations are required to establish the place of EGR/3-way catalyst engines for future low-emissions cogeneration. For example, would a "spark plug amplifier" be necessary for a large engine of this type? If so, a supply of both gas and air to a prechamber would probably be necessary.

2.2 Dual Fuel Engines

The ability to operate on gas or diesel fuel remains an attractive option for some applications and the high pressure direct gas injection (HPDGI) engine is an exciting entry into this field, offering full diesel bmep and efficiency capabilities on gaseous fuels with no knock limitation [3, 4, 5, 6]. This type of engine cannot be expected to achieve really low NOx emissions, however, and the cost, complexity and power demand of the high pressure gas compression and injection systems are significant. With exhaust aftertreatment, however, this engine is expected to become an important power unit in cogeneration plants where shaft efficiency is a high priority. It makes much lower demands on the aftertreatment plant than a straight diesel engine.

The traditional type of dual fuel engine with premixed gas and air combustion is closely analogous to the SI lean-burn engine in its low emissions potential and knock-limited power capability. To achieve the best low-NOx and low-particulate performance the pilot diesel fuel quantity should be reduced to a level below that which the normal diesel fuel injection system can successfully deliver. Like the prechamber of an SI engine, the fuel burned as the rich ignition source ought to be below 2% of the full load engine fuel energy input and so a true low-NOx dual fuel engine must have a specially matched injection system for pilot fuel delivery. This small volume system would not be able to give full power operation on diesel fuel though it must be capable of starting and operating the engine without gas in the true dual fuel case. However, a diesel-ignited gas engine, with no

TABLE 1

MAIN TECHNOLOGIES FOR GAS ENGINE DEVELOPMENT

• Better Combustion Control (Prechambers and Mainchambers)	• Better Combustion Models
• Mixture Quality	• Air/Fuel Ratio Sensing
• Ignition System Durability at High BMEP	• Knock Sensing
• Control of THC Emissions	• Control Philosophies
• Dual Fuel (Diesel/Gas) Optimisation	• High Pressure Direct Gas Injection (HPDGI)
• Improved Catalyst Formulations for Methane Conversion	
• Gas Quality Sensing/Control	• Lubricants (Standards)

TABLE 2

TYPICAL POTENTIAL FOR VARIOUS ENGINE TYPES AT 1 MWe

	DIESEL ENGINES		GAS ENGINES (1995)			
	CURRENT DI	1995 DI	LEAN BURN PCC + CONTROLS	λ=1 + TWC*	λ=1 + EGR + TWC*	HPDGI
BMEP bar	18	20	14	8	12	20
RELATIVE EFFICIENCY %	100	97	85	77	80	98
EMISSIONS POTENTIAL g/Nm^3 @ 5% O_2						
NOx	2.0	1.5	0.3	0.15	0.1	1.1
THC (incl. Methane)	0.18	0.1	2.0	0.6	1.0	0.7
CO	0.2	0.23	0.5	0.5	0.85	0.25
PARTICULATES	0.18	0.09	0.02	<0.02	<0.02	0.04

* Engine out emissions quoted, except where 3-Way Catalyst (TWC) is indicated

capability to deliver power on liquid fuel only, is a realistic alternative to the spark ignited engine as a low emissions gas-fired power plant. The reliability and durability of the diesel injection system would be expected to far exceed that of today's spark ignition systems but the diesel-ignited engine would have to retain a compression ratio adequate for its operation. This might limit its application to gaseous fuels with the highest anti-knock characteristics only. The specially matched pilot injection may be made directly into the main cylinder or into a separate prechamber. The future availability of high ignition quality low sulphur diesel fuel is assured for automotive purposes and the added cost and burden of supplying the second fuel for ignition purposes is likely to be acceptable for cogeneration in the future scenario of more complex engines meeting more stringent regulations.

Table 1 lists the technologies to be further developed and Table 2 indicates emissions expected for various engine systems.

3. EXHAUST AFTERTREATMENT FOR IC ENGINES

There is little doubt that as pollution control becomes more comprehensive, no IC engines without aftertreatment will be able to comply with regulations in certain areas. The further development and control of catalytic processes remains a key factor in the future of gas engines for cogeneration, therefore. The need for improved Methane conversion has already been mentioned. To ensure satisfactory long term effectiveness in service, the demands on the aftertreatment systems must be kept at levels which are below their maximum possible conversion capabilities. This means that reductions in engine out emissions will still be needed.

SCR for NOx reduction appears set to move into a period of expansion, with applications to smaller plants becoming more common. The low-sulphur content of most natural gas fuels removes one headache, but the technology, together with its combination with oxidation catalysts and heat recovery equipment, continues to develop. The possibility of unintentionally releasing large amounts of Ammonia into the environment must be minimised; these plants are likely to operate for 20 years and more. Catalytic NOx removal without a reducing agent is the subject of much promising investigation but it is too early to say how successful this will be.

For the 3-way catalyst, the optimisation of formulations for natural gas is still an objective and further experience with EGR is needed.

Non-catalytic methods of aftertreatment would be expected to depend largely upon high reaction temperatures and the development of processes for use in combination with afterfiring is likely.

Aftertreatment processes continue to require better engine air/fuel ratio control with increased reliability. More difficult, however, will be the need to ensure that aftertreatment systems do not create new problem pollutants. This is an area of some concern and of course not all such unwanted products are organic. NH_3 and N_2O are amongst the "simple" undesirable materials emanating from engine catalytic systems.

4. ROUTES TO FURTHER IC ENGINE IMPROVEMENT

The IC engine is a highly developed and mature product. Although no major leap forward in performance or low emissions is available and no revolutionary new concept is expected, there are further improvements to be made. Despite the maturity and experience of the industry, however, the routes to achieve the necessary improvements to pollution control and efficiency are not obvious. If we are to further reduce the NOx from a spark-ignited gas engine, many questions arise as to the best route forward. Examples are:

If charge dilution is employed, how much should there be? Is it better to aim for a rather retarded ignition timing with some dilution or to have as much dilution as possible and then to advance the timing for efficiency? What about the compression ratio? How does it trade-off against ignition timing in the NOx/efficiency/knock equation? What combustion chamber features give the lowest THC emissions whilst retaining acceptable dilution tolerance and cycle-to-cycle repeatability?

These questions and many more like them are not easily answered because of the complex interaction of many factors in the combustion of gas/air mixtures in engines. For example, the combustion rate for any one combustion chamber and gas composition is affected by engine speed and load, ignition timing, charge trapped temperature and pressure, compression ratio, air/fuel ratio and coolant temperature. Many of the same factors also influence the effective start of combustion, so the timing as well as the rate are affected by these variables. Therefore they affect the NOx emissions and brake thermal efficiency directly and the THC and other emissions more indirectly.

In the light of such influences the setting of a design and development strategy for future engines is difficult. Other factors affect the emissions/efficiency trade-offs too. Many of these effects can be calculated and therefore engine performance modelling should form a powerful source of guidance. In practice, however, this is not the case with most engine makers. The reason for this is that a model of premixed combustion is not yet available which is adequate for development use. Such a model should take into account the effects of combustion chamber geometry and size because these influence turbulence, prior to and during combustion. The phenomenon of detonation should also be modelled and the influences of fuel composition upon the onset of detonation incorporated. The factors governing the growth and development of a viable flame kernel from the spark plug gap are not yet well enough understood to enable this phase of the combustion process to be reliably modelled. Finally, the influences of pressure, temperature, turbulence, air/fuel ratio, gas composition, residual gases, etc. on the flame development and burning rate have to be predicted. Then the NOx and efficiency could be calculated. (The effects of crevices and surface features on Hydrocarbon emissions would also be desirable for such a model). The absence of such a comprehensive model for gas engine combustion largely explains the limitation in the use of engine cycle simulation in emissions reduction programmes.

Where a large amount of test data and experience is available, it is possible to set up some empirical models for the combustion phase. This immediately places limits on the applicability of trends predicted by such a model, but when used with engineering care and experience this type of model can be useful [7]. Based on this careful use, sufficiently good representations of combustion are possible to enable efficiency and NOx to be predicted. One of the authors uses this type of modelling, including NOx calculations developed on from Reference 9 and an empirical knock assessment method. The model also realistically slows down the burning rate as the air/fuel ratio increases, based on an extensive database and experience. Figure 2 shows an example of an output from such a study in which the importance of burning rate on the NOx/efficiency trade-off was being evaluated. The four separate graphs show this trade-off for four different Start of Combustion timings (SOC). In each graph there are three curves, each for a different combustion rate, described as Fast, Medium or Slow. The air/fuel ratio (Lambda) is then varied from 1.0 (stoichiometric) to 1.7 (very lean) for the three basic burning rates to form each curve. The estimated knock boundary is also shown. In this example the manifold conditions were kept constant so the power output is different for each point. Another set of curves would be required for different compression ratio or manifold conditions.

The results from such simulations thus have to be treated with caution and conclusions drawn with a leaven of experience. The guidance given can be cost effective but the need for better models is clear. Improved engine simulation codes which are usable for development purposes are becoming available [9] and Computational Fluid Dynamics (CFD) is seen as one technique of particular importance so that flows, mixing, turbulence and flame development can be optimised [10]. By such means, the IC engine will continue to improve in performance and environmental friendliness.

The Fuel Cell will be firing at a target which is still moving in the 21st Century.

5. OTHER CONTENDING COGENERATION PRIME MOVERS

Before considering Fuel Cells, it is important to ask whether the gas-fired reciprocating engine will remain the main incumbent which is to be superseded. The

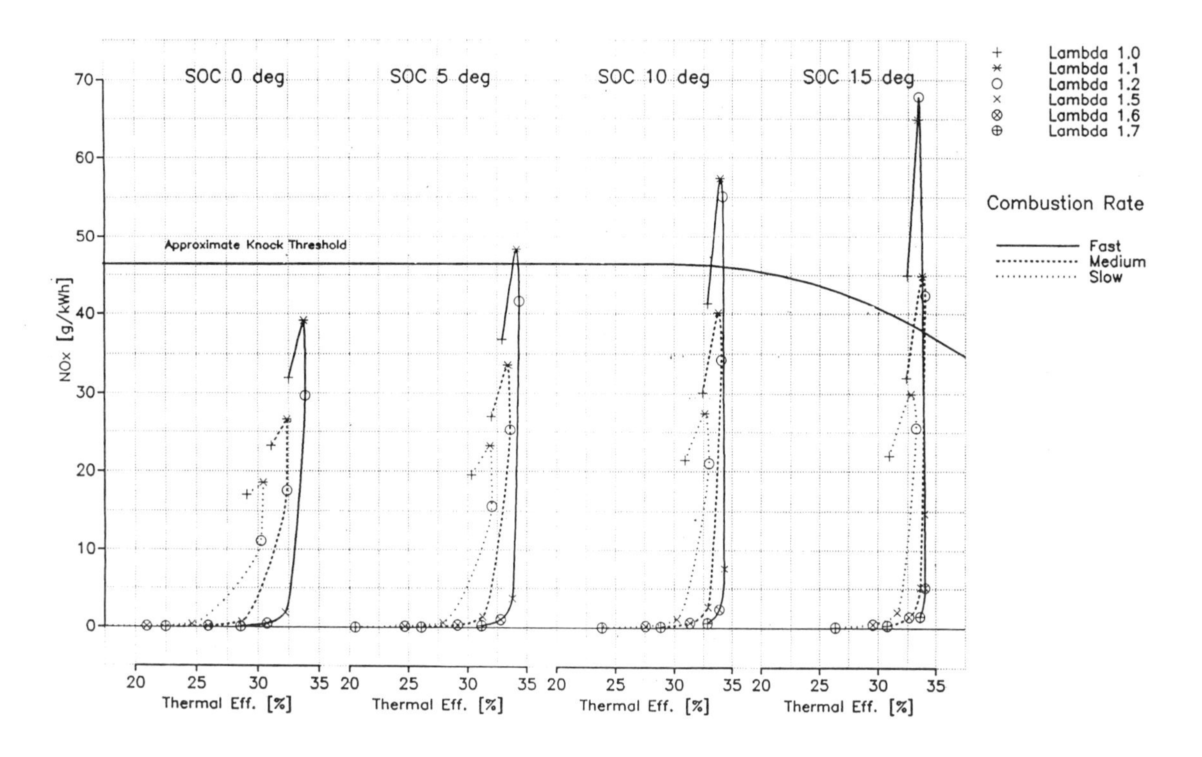

Figure 2 - Computed NOx/Efficiency Trade-offs for 4 Timings and 3 Combustion Rates

authors view the Gas Turbine and the Stirling engine as the main competitors and other serious contenders are not foreseen, although the high efficiency of the Diesel engine must not be ignored. The Gas Turbine is expected to become more competitive in the upper end of the power range considered in this paper. When burning Natural Gas its potential advantages in time to overhaul and reliability can be realised. It is receiving large investment but the extent of its inroads into the market cannot be clearly predicted at present, especially as the size of the market itself is not reliably known. The Stirling engine does not appear likely to make progress into this market in the short to medium term. The limited effort being put into its development seems to confirm this. The Stirling engine cannot be thought of as a kind of extension of IC engine technology and its future development to levels where it realises some of its theoretical potential would be extremely expensive.

6. THE FUEL CELL

6.1 Introducing the Fuel Cell

As a replacement for the ubiquitous reciprocating engine the Fuel Cell, which has been around for about 150 years, is beginning to throw down a serious challenge. Improvements in materials technology, for example, now mean that concepts which appeared confined to the laboratory ten years ago are currently being pushed hard towards pre-commercial demonstration. If this is successful, then the gas engine may have a serious contender for both power generation and some transport applications.

Firstly, it is important to understand some of the fundamental principles and limitations of conventional fluid power plant. The Second Law of Thermodynamics leads us to the Carnot Cycle and then to the more practical Rankine, Joule and Diesel Cycles, among others. The overall efficiency of such prime movers is limited both by basic physical laws and by engineering considerations, the latter mostly relating to operating pressure and temperature.

The perfect Fuel Cell is an electrochemical energy converter which would have the same efficiency as the Carnot cycle engine [1]. The 'real world' effects, particularly of temperature, which sap the performance of the 'ideal' engine, affect Fuel Cells much less, however. Some, for example, can approach to within a few percent of their maximum theoretical performance.

Reciprocating engines have benefited from technological advance over a very long period and gas turbines too have had many years of development. Used in the combined cycle mode, the latter will develop much further yet. The Fuel Cell should therefore be viewed against this background. Its ultimate markets will probably be more oriented towards the small reciprocating engine cogeneration scene, but it will be extremely demanding of technology. It offers two potential advantages: generation efficiencies higher than are likely from existing alternatives and very significant reductions in NOx emission levels - virtually zero in some cases. A related issue is the part load performance, which holds up well, to perhaps 20% load. In this context, the reciprocating engine is less good but is significantly better than the gas turbine. Thus, for load following applications, the Fuel Cell's characteristics are very appropriate.

The countries leading these developments are the United States and Japan and strong programmes are now beginning to emerge in Europe. Very significant investments have been involved: $360m over 10 years for PAFC's alone in the United States. The pull of environmental awareness and energy conservation should continue to provide such investment.

TABLE 3
FUEL CELL TECHNOLOGY

	FIRST GENERATION			SECOND GENERATION	
TYPE	AFC ALKALI FUEL CELL	SPFC SOLID POLYMER FUEL CELL	PAFC PHOSPHORIC ACID FUEL CELL	SOFC SOLID OXIDE FUEL CELL	MCFC MOLTEN CARBONATE FUEL CELL
OPERATING TEMP.	< 100°C	< 100°C	< 200°C	900/1000°C	650°C
APPLICATION	Power	Power	Power/Cogen (LP Steam)	Cogen	Cogen
PLANT POWER AND OPERATIONAL EXPERIENCE	Aerospace Well Established	Aerospace Well Established	1 MW (Milan) 11 MW (Tokyo) (Target of 1900 MW in Japan by 2000)	100 kW (1994 target Japan) (1998 Target Europe) 10 kW; 30000h operation 3 kW; 10000h operation	100 kW (1991 target USA) 2 MW (1993 target USA) 1 kW; 20000h operation 10 kW; 5000h operation
TYPICAL EFFICIENCY	Low but 73% on $H_2 + O_2$ mixture	Low	Circa 40-50%	Circa 55% (Rises to 65% with steam turbine)	Circa 55%
INSTALLED COST			$4-6000/kW $2000/kW (1995) $1000/kW (2000)	$1000/kW target	$1000/kW target
COMMENTS		Costly platinum catalyst. CO will poison catalyst, fuel preprocessing required.	Costly platinum catalyst. Fuel preprocessing required.	Simple internal fuel processing possible. 85% fuel conversion. Catalytic exhaust oxidation. SO and CO tolerant.	Simple internal fuel processing possible. 85% fuel conversion. Catalytic exhaust oxidation. SO and CO tolerant.

6.2 Fuel Cell Concepts

Table 3 illustrates a range of viable Fuel Cell technologies. The Fuel Cell is an electrochemical reactor producing power and heat. It is fundamentally limited in size and is extremely delicate, usually either of tubular or planar construction, of overall dimensions less than a metre. Some prototypes are only a few centimetres across. Materials of construction include carbon, polymers and ion conduction ceramics. Individually, each Cell will deliver something less than 1 V and between 10 and 100 W, perhaps eventually 1 kW. Cells are therefore stacked and manifolded together to produce modules at whatever level of power is required. This modularity is one of its fundamental advantages. For a 100 MW power station, the system would incorporate both gas and steam turbine plant, to form a 'Composite', with the Fuel Cell effectively replacing the gas turbine combustor.

Figure 3 illustrates the basic principle of the Fuel Cell. In terms of achievement to date, the Alkali (AFC), and Solid Polymer Fuel Cells (SPFC) have played a vital role in the exploration of Space and have also seen applications in Submarines and Submersibles. Both run at low temperature (<100°C) thereby offering limited potential as cogenerators and are also costly - one reason being the requirement for expensive (platinum) catalyst materials. A more recent development, the Phosphoric Acid Fuel Cell (PAFC), has now reached the pre-production stage and Demonstration plants of American/Japanese design are in operation. These will shortly be joined in Europe by a 1 MW plant in Milan. A large 11 MW plant is currently under construction in the USA for Tokyo Electric (TEPCO).

The PAFC (Figure 4 shows an Westinghouse unit) operates at slightly higher temperatures (~200°C) at efficiencies comparable only with conventional plant. Its primary advantages are its low NOx emissions and modularity, lending itself primarily to small scale distributed cogeneration applications using hot water or LP steam. One of the problems of the PAFC is first cost, which presently stands at $4000-$6000/kW and this must fall significantly before the commercial stage is reached. Platinum loadings are again an issue.

Two other types of Cell, considered to be second generation, are currently at an R & D stage and these are the Solid Oxide (SOFC, one concept is shown in Figure 5) and Molten Carbonate (MCFC) variants. SOFCs and MCFCs offer higher Cell efficiencies in their own right, run at much higher temperatures and are therefore better for cogeneration.

Figures 6 and 7 give an indication of a simple and a Composite system respectively. These are based on the SOFC. The Fuel Cell stack alone would offer efficiencies towards the upper fifties with useful waste heat. Its modularity could lead to capacities from around 5 kW upwards. The incorporation with turbine plant could raise generation efficiencies to the mid sixties, but would dictate a minimum capacity of several tens of megawatts. A tubular design of SOFC is available pre-commercially, but the most desirable planar variant is some years away.

SOFC's are targeted to demonstrate 100 kW by 1994 in Japan and 1998 in Europe. A recent ambitious USA initiative by the APPA aims to demonstrate MCFC technology at 100 kW by the end of 1991 and 2 MW by 1993.

The fundamental advantage of the high temperature Cells is that there is no longer a requirement for a noble metal catalyst, suggesting lower ultimate costs. In

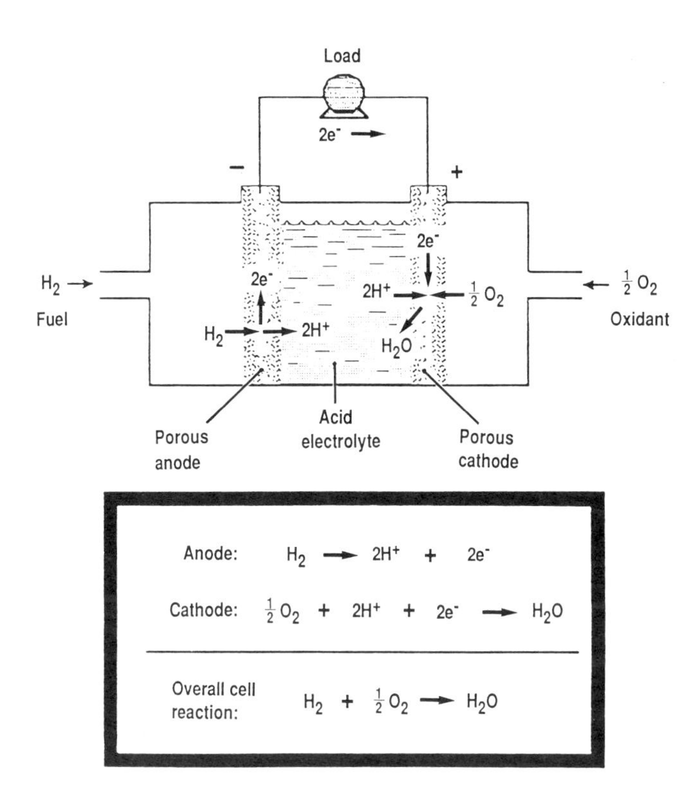

Figure 3 - Principle of Operation of a Fuel Cell

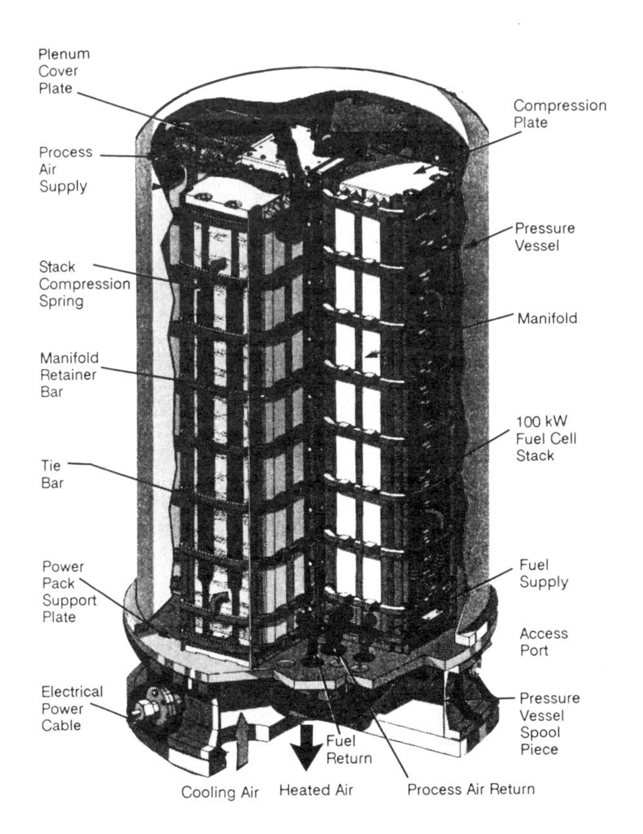

Figure 4 - Westinghouse 500-kW Fuel Cell Module

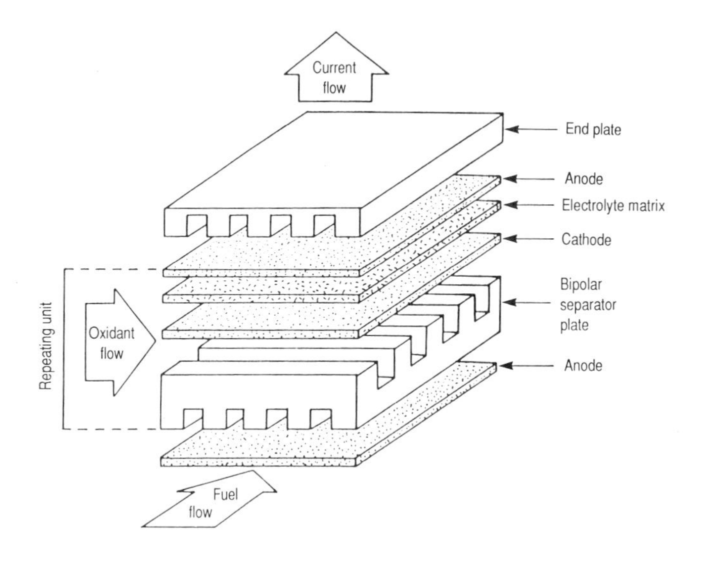

Figure 5 - Typical Solid Oxide Fuel Cell

the context of hazardous or strategic materials, clearly phosphoric acid must be treated with due caution. The depleted fuel streams from the high temperature Cells pose a potential hazard since they contain about 15% residual fuel. Platinum is not expected to become too scarce and the rare earths used in the ceramics for high temperature cells are not truly "rare".

6.3 Fuels for Fuel Cells

The Fuel Cell is unique in that the fuel and oxidant streams remain separated throughout and there are therefore two exhaust streams from the stack itself. Subsequently, the flows may be combined for catalytic combustion, creating further heat for the user and/or feed stream preheating.

The ideal fuel and oxidant are hydrogen and oxygen respectively, although air is clearly preferable. Hydrogen is chemically the ideal fuel but Natural Gas, a more likely feedstock, will reduce efficiency somewhat. For PAFCs and other low temperature variants, Natural Gas must be reformed upstream of the Cell to yield a hydrogen/carbon monoxide/carbon dioxide fuel stream. External stream reforming is used with carbon monoxide reduced to a minimum through a second stage water-gas shift reaction. This is necessary since carbon monoxide has a poisoning effect on the anode catalyst, particularly with the SPFC.

The higher temperature Cells offer the prospect of reforming within the Cell itself, which would greatly simplify the overall plant design by allowing operation directly on Natural Gas or gasified coal. Internal reforming brings additional problems for the Cell, particularly carbon deposition on the anode and the process must be carefully designed to alleviate this problem. The higher temperature Cells are much more tolerant of sulphur or carbon monoxide content in the feedstream than the other types.

6.4 Fuel Cell Costs and Market Entry

To estimate a target price at which Fuel Cells will become financially viable it is assumed that there will be a financial value placed on the environmental

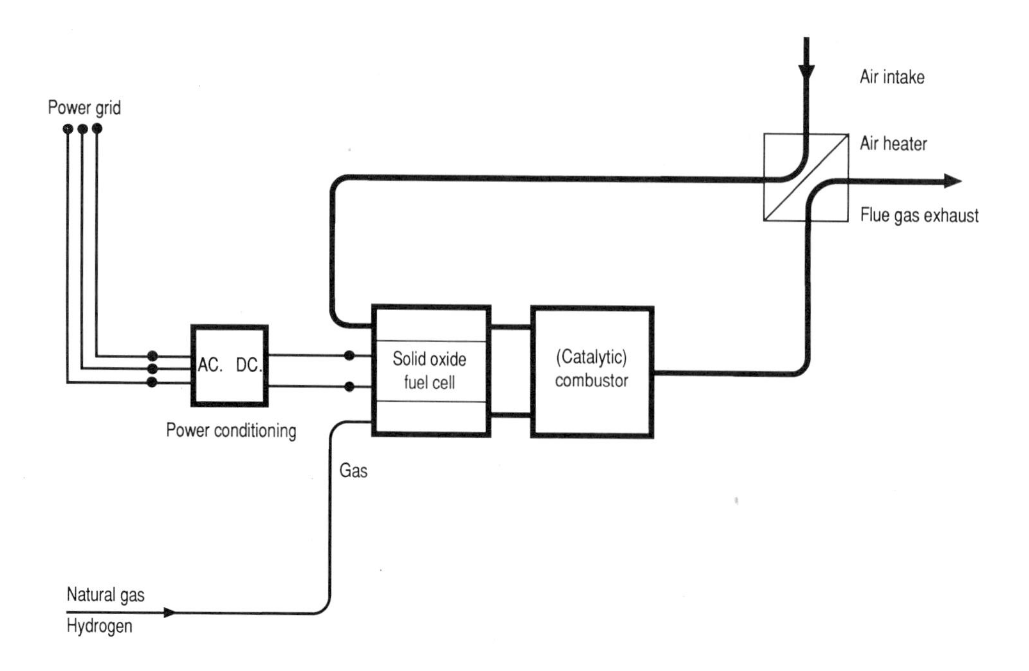

Figure 6 - Simple Solid Oxide Fuel Cell

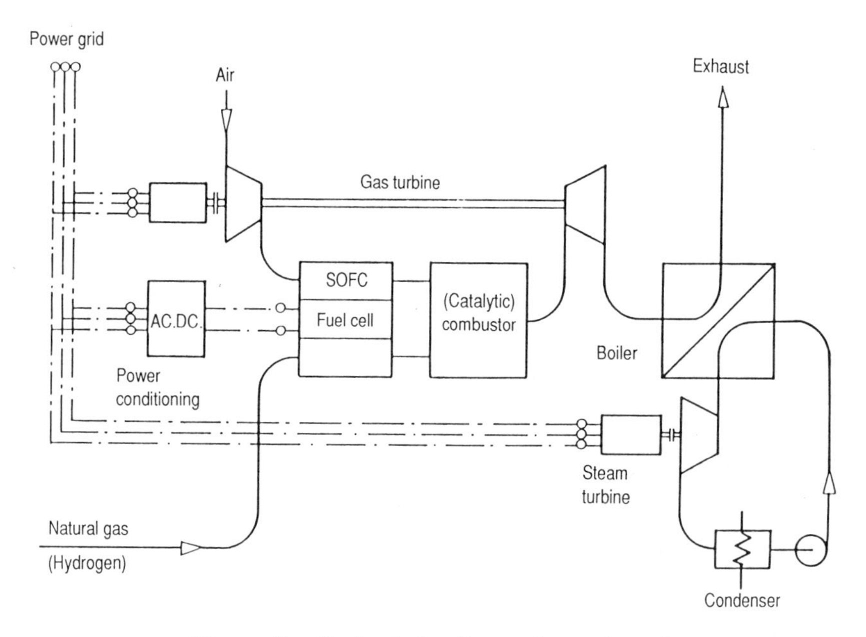

Figure 7 - Typical Gas/Steam Composite Plant

benefits. Ultimately, the cost of producing electricity should be the critical factor but, for successful market entry, the Fuel Cell will have to show some 'margin of advantage'. For larger units the cost of electricity will be the main issue, but for small scale plants first cost will be the key.

It is considered that a general target of £1000/kW must be achieved for successful market entry. However, this target will be affected by the particular sector, the country and the "cost of electricity" argument. PAFCs are targeted to be down to $2000/kW by 1995 and $1000/kW by the year 2000. It is forecast that the high temperature variants will eventually be able to achieve $1000/kW or less. Market introduction and penetration will depend upon the achievement of satisfactory unit costs and it is not really possible to predict the ultimate market penetration, except perhaps in Japan. There, with high energy prices, there is a real commitment to install 1900 MW of PAFC capacity by the year 2000. With this achieved, and with the country's manufacturing and cost reduction skills, there is a strong possibility of Japanese plant entering the world market in an advantageous and highly competitive position.

7. CONCLUSIONS

The Natural Gas-fired reciprocating engine will form the main target at which Fuel cells will aim in the low emissions 50-3000 kWe range. Diesel and Gas Turbine plants continue to develop and will remain strong competitors.

The price of gas engines which meet the next Century's environmental needs will be higher than that of today's engines because of added sophistication, controls and aftertreatment. Cogeneration package prices will probably fall slightly, however. If Fuel Cells achieve the $1000/kWe target (1989 prices) they will certainly enter the reckoning for potential purchasers of low emissions high efficiency cogeneration.

Further improvements to reciprocating gas engines require adequate investment in technology. Gains will come both from engine improvements and from the further development of aftertreatment systems. Radically new IC engine concepts are unlikely to appear but subtle changes can be expected.

Particular areas where more work is required for gas-fired engines are improved prechamber technology with size reductions, further evaluation of the potential for EGR with 3-way catalysts, control of Methane emissions from lean-burn engines and the optimisation of dual fuel engines with minimum pilot fuel quantities. The high pressure direct gas injection dual fuel engine has considerable potential for some applications, mainly at the higher unit powers.

Engine simulation involving much empirical modelling is currently useful to assist for low emissions engine development and problem solving. Much care and experience is required in achieving reliable results, however. Better codes must include more comprehensive combustion models.

There should be a booming market in aftertreatment equipment for gas engines as the end of the Century approaches. 3-way, oxidising and SCR catalysts will all find opportunities and these will encourage the use of powerful engine management systems. The development and cost reduction of aftertreatment systems is greatly in the interests of the market for reciprocating gas engines.

The Fuel Cell is a powerful contender for the environmentally conscious power generation market that will be well established by the end of the century. It is not yet certain which of the various types is likely to be the most successful, but the level of international commitment and achievement to date suggests that there will be an ultimate success, urged on by the demand for cleaner and more efficient power generation.

The Fuel Cell has the fundamental advantages of modularity, simplicity, silence of operation and potential high performance. The questions of reliability and maintenance costs still have to be proven. Cells have run for 30,000h continuously although performance on Natural Gas under load following applications needs demonstration. The Cell performance does deteriorate with time but guarantees to 40,000h should be forthcoming for PAFCs.

A cost target of $1000/kW is considered realistic to enable the Fuel Cell to become commercially viable. It is predicted that PAFCs will reach this target by the year 2000 and that high temperature Cells will eventually achieve competitive prices. The pace of development in Japan suggests that that country may establish a technological and commercial lead in the field.

ACKNOWLEDGEMENTS

The Authors wish to thank the Directors of their respective Companies for permission to publish this review and their colleagues for much support.

REFERENCES

1. Hundleby, G. E. and Thomas, J. R.
"Low Emissions Engines for Heavy Duty Natural Gas-Powered Urban Vehicles - Development Experience".
SAE 902068 1990.

2. Quader, A.
"Why Intake Charge Dilution Decreases Nitric Oxide Emission from Spark-Ignited Engines".
SAE 710009 1971.

3. Einang, P. M. et al.
"High Pressure, Digitally Controlled Injection of Gaseous Fuel in a Diesel Engine with Special Reference to Boil-off from LNG Tankers".
CIMAC D11.5 1983.

4. Miyake, M. et al.
"The Development of High Output Highly Efficient Gas Burning Diesel Engines".
CIMAC D11.2 1983.

5. Schnor, O.
"MAN-B&W Dual Fuel Engines - 4 Stroke".
Symposium on Gas Fired Cogeneration Technology, Aarhus 1987.

6. Gros, S. and Vestergren, R.
"High Pressure Gas Technology and Emission Control for Modern 4-Stroke Diesel Engines".
I.Mech.E. Seminar on Gas Engines and Cogrneration, Solihull 1990.

7. Blumberg, P. N.
"Selecting and Applying Engine Simulation Methods to Achieve Engineering Objectives".
ASME ICE Vol. 9 1989.

8. Nightingale, D. R.
"A Fundamental Investigation into the Problem of NO Formation in Diesel Engines".
SAE 750848 1975.

9. Morel, T., Keribar, R. and Blumberg, P. N.
"A New approach to Integrating Engine Performance and Component Design Analysis through Simulation".
SAE 880131 1988.

10. Johns, R. J. R. and Jones, P. M.
"Computer Modelling of the Flow in a Lean Burn Natural Gas Engine".
I.Mech.E. Conference - "Computers in Engine Technology".
Cambridge, UK 1991

11. Appleby, A. J. and Faulkes, F. R.
"Fuel Cell Handbook" 1989.

ICE-Vol. 15, Fuels, Controls, and Aftertreatment
For Low Emissions Engines
ASME 1991

A STUDY OF ELECTRONIC ENGINE CONTROL SYSTEM STRUCTURES FOR LEAN-BURN, NATURAL GAS OPERATION IN A SPARK IGNITION ENGINE

Ronald S. Patrick, Alan R. Eaton, and J. David Powell
Department of Mechanical Engineering
Automotive Controls Laboratory
Stanford University
Stanford, California

ABSTRACT

Various engine control systems using production and laboratory sensors, actuators, and control algorithms were assembled and tested on a natural gas fueled SI engine. Control algorithms included feedforward, exhaust oxygen sensor feedback, and cylinder pressure sensor feedback for AFR and MBT control. The goal was to provide fast, potentially cycle-to-cycle, lean-burn control. Advantages of timed injection versus continuous injection, injection phasing control, and electronic throttle actuation for lean limit extension and cycle-to-cycle control were found.

BACKGROUND

In recent years, aspirations for improvements in internal combustion engine efficiency, emissions, drivability and fuels usage have increased. Natural gas, because of its availability and high knock resistance, shows promise in reducing our dependance on liquid hydrocarbon fuels.

The use of a new fuel should be met with a consideration of the suitability of sensors, actuators, and control algorithms (i.e. the engine control system (ECS)) currently used with large-volume, production gasoline engines as well as those sensors, actuators, and algorithms still in the laboratory.

To this end, an experimental study was performed in which production and laboratory components and concepts were used to assemble various control systems for a lean-burn, natural gas fueled spark-ignition (SI) engine.

The Engine Control Problem

The essence of the engine control problem for vehicular applications is to minimize fuel consumption, keep emissions below mandated limits and provide acceptable drivability, while operating over a range of engine torques and speeds. In most gasoline-fueled engines, the driver's command for an acceleration, deceleration or maintenance of constant vehicle speed is realized as air throttle position. The ECS responds to this command primarily by adjusting the control variables of fuel pulse width and ignition timing. With a 3-way catalyst, the overall engine control solution is to maintain an air-to-fuel ratio (AFR) close to stoichiometric and ignition timing at minimum advance for best torque (MBT). This solution has allowed current standards to be met, albeit with some exhaust gas recirculation (EGR) and ignition timing retarded from MBT.

Lean-Burn Engine Control

The principal benefit of operating an SI engine at AFRs lean of stoichiometric is an increase in thermal efficiency. Unfortunately, engine NO_x production peaks at lean AFRs and no suitable reducing catalyst currently exists for lean operation. However, should the engine be able to operate sufficiently lean, NO_x output may be reduced to an acceptable level. This places the AFR closer to the lean limit of the engine. Maintaining stable operation at these AFRs is difficult and for some engines is impossible. As a result, the challenge of AFR control for a lean-burn engine is not to maintain catalyst efficiency, but to avoid lean misfires.

The most common form of ECS for lean-burn operation is one in which the driver controls the air throttle. With this system, the overall engine control solution is to regulate a lean AFR and ignition timing at MBT. Variants of this system use EGR and/or ignition timing retarded from MBT to suppress NO_x or knock.

A second form of lean-burn ECS takes control of the air throttle away from the driver and assigns it to the engine control computer (drive-by-wire). In this structure, the driver's command controls the rate of fuel delivered to the engine. The AFR is always lean except when the driver's command is at its maximum. The ignition timing is most often at MBT. The air throttle and EGR rate are controlled by the ECS so as to maximize fuel economy and minimize emissions at each torque/speed condition. In some engines, EGR may be used in larger quantities than necessary to suppress NO_x so as to improve fuel efficiency largely through reduced pumping losses and reduced heat transfer. Advantages of this form of lean burn ECS are improved torque response to the driver's command and improved control system stability (Stivender, 1978).

Scope of this Study

This study was restricted to lean-burn control solutions of the first form; those in which the driver controls the air throttle.

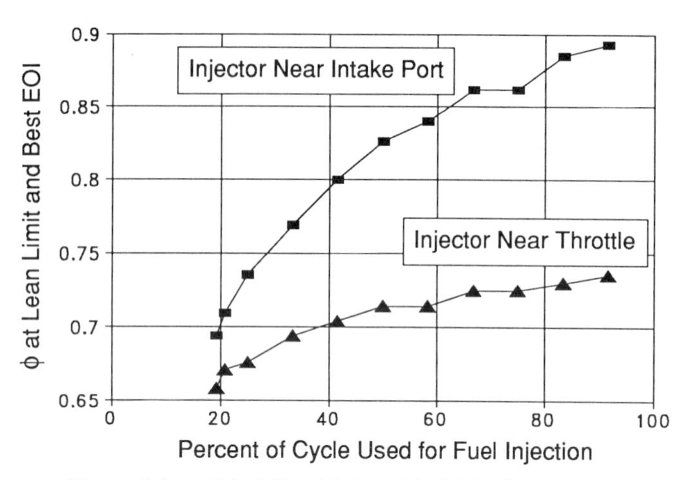

Figure 1: Lean Limit Sensitivity to Fuel Injection Duration (1000 RPM, 40°TVO)

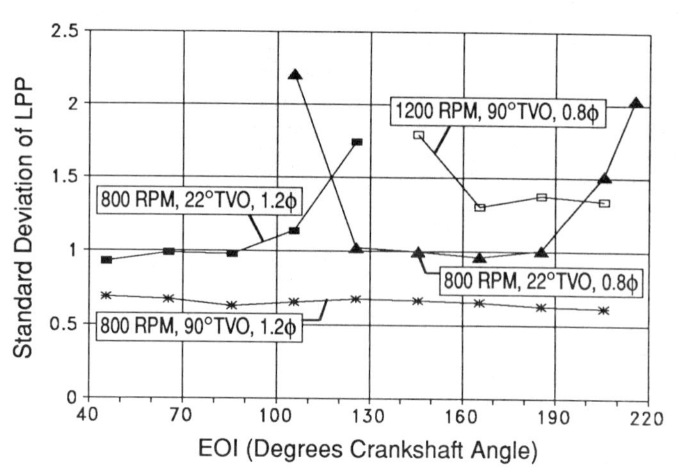

Figure 2: Engine Stability Sensitivity to Fuel Injection Phasing (Injector Near Intake Port)

In addition, because the amount of EGR required to reduce NO_x emissions is engine dependent, the effect of this parameter was not studied (i.e. no external EGR was used). Thus, the lean-burn control problem studied was reduced to one common to all engines of this type; how to maintain control of AFR close to the lean limit and at MBT.

CONTROL ALGORITHMS

For a given engine configuration (combustion chamber, intake port, etc), electronic control can influence lean-burn engine operation in three ways:

1. By providing additional control variables which can be used either to extend the engine's AFR, EGR ratio and ignition timing limits of operation or to improve its performance at a given operating condition.

2. By allowing a means by which the AFR, EGR ratio and ignition timing may be regulated to their targets over the engine's range of operation. This range may be defined in terms of operating parameters and engine parameters. Operating parameters include: manifold charge pressure, manifold charge temperature, exhaust pressure, engine temperature, AFR, EGR ratio, humidity, fuel characteristics, and engine speed. Engine parameters include: displacement, compression ratio, camshaft profile, intake system flow coefficients, EGR valve flow coefficient, fuel injector flow coefficient, and exhaust system flow coefficients. Engine parameters will vary for a given configuration due to manufacturing variations and degradation.

3. By providing a means to determine new AFR, EGR ratio and ignition timing targets so as to reoptimize the control to compensate for engine and fuel variations.

Control Variables

Most natural gas fueled engines in use today use continuous flow fuel delivery systems. With these systems, the basic fuel rate to the engine is determined by the air flow through a venturi with modifications (trims) of this rate made by controlling the pressure upstream of a flow orifice or by varying an orifice's flow area. Since most production gasoline-fueled engines use timed flow, it is of value to compare the performance of these two fuel delivery systems for a gaseous-fueled engine.

Figure 1 shows experimental results from a single cylinder CFR engine (see Appendix A for engine, sensor, and actuator specifications) using natural gas as a fuel. This engine has an electronic fuel injector whose pulse width and location of pulse width with respect to crankshaft position (injection phasing) are controlled. In addition, the gas supply pressure upstream of the injector is controlled. Thus, combinations of pulse width and supply pressure can be used to deliver the same mass of fuel to the engine. The combination for which the pulse width is the duration of the engine cycle is essentially a continuous flow system.

In this figure, the engine speed and the air throttle position (TVO) were held constant. Various fuel pulse widths (shown on the abscissa as percentage of engine cycle) were set and at each percentage the supply pressure and injection phasing were adjusted to achieve the leanest stable operation (shown on the ordinate as equivalence ratio, ϕ = stoichiometric AFR/actual AFR). Three misfires per thousand cycles defined the lean limit. Although a fully continuous flow system was not tested, there was a definite trend showing the advantage of a timed system in extending this engine's lean limit of operation. Certainly, the physical characteristics of our particular engine affected the trend. So to study the effect of an engine configuration variation on the trend, the fuel injector was placed in two locations: the first at the intake port and the second just downstream of the air throttle. In both locations, the trend remained the same. For this reason, the engine was operated using a timed fuel injection system with the highest injection pressure practically possible.

In most production, stoichiometric target, gasoline-fueled engines using timed fuel injection, the phasing of the pulse width is not a control variable. In these systems, the end of injection (EOI) or the beginning of injection (BOI) is fixed with respect to crankshaft angle. This fixed value is principally chosen so as to reduce HC emissions from unvaporized fuel. In contrast to the stoichiometric, gasoline-fueled engine, little work has been done to illustrate the value of phasing as a control variable for lean-burn, gaseous-fueled engines. Because gaseous-fueled engines do not have unvaporized fuel and because lean-burn engines may operate over wider ranges of AFRs than stoichiometric target engines, a study was made of the sensitivity of engine operation to this parameter.

At combinations of torque, speed, and AFR spanning the test engine's range of operation, the EOI (and hence injection phasing) was varied. Engine stability at each EOI was measured in terms of the standard deviation of the location of peak cylinder pressure for one thousand cycles. Figure 2 contains representative results.

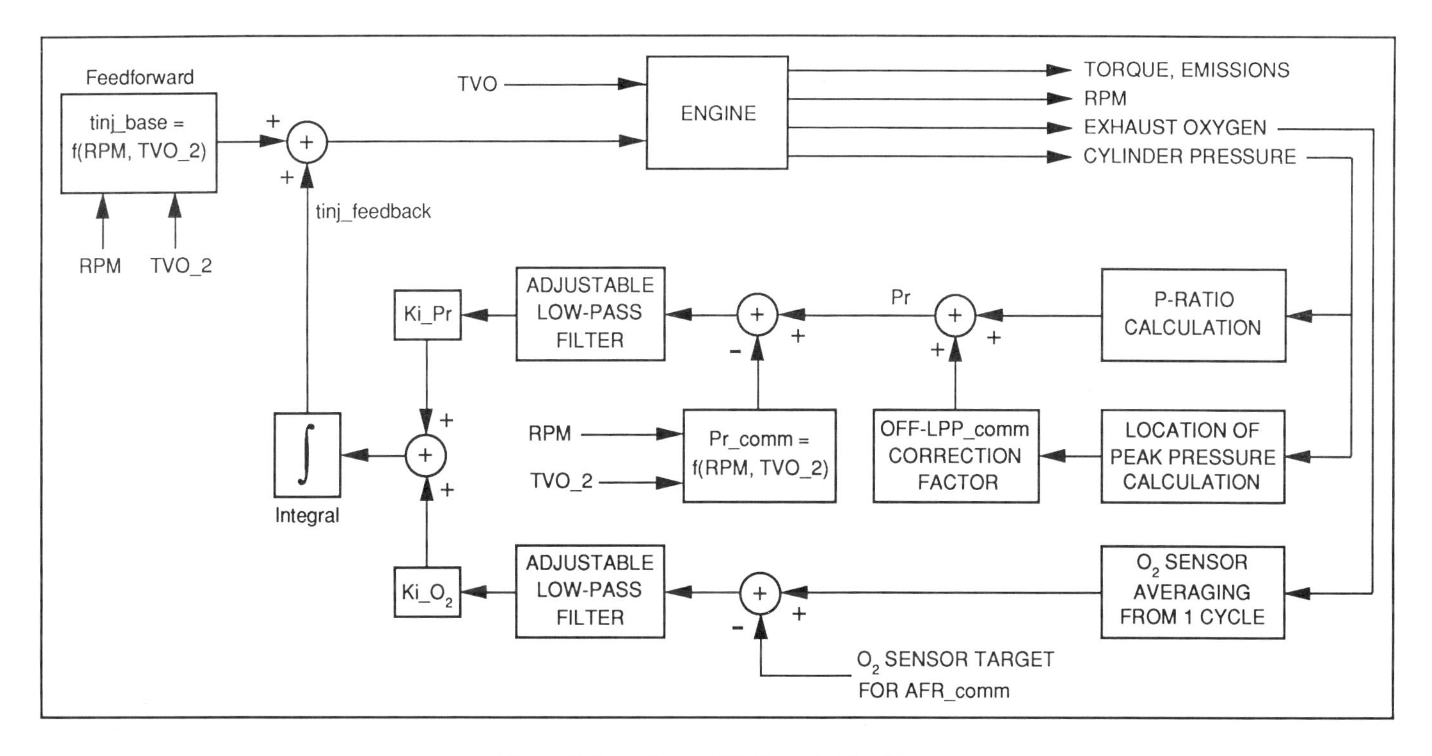

Figure 3: Lean-Burn AFR Control Algorithm

Clearly, the control parameter of injection phasing is a powerful one in extending this engine's AFR range of operation. The extent to which an engine's configuration (i.e. intake port, intake valve, and combustion chamber shape) influences this sensitivity should be explored.

For this study, the control variables were limited to fuel pulse width and ignition timing. The fuel injection phasing was fixed for the remainer of the tests at a "good" value (EOI=185°, see Figure 4) for the lean AFR target chosen. The phasing was not varied because, as will be observed, the phasing should be made with consideration of air throttle motion.

AFR Control Logic

The logic for AFR control must provide a means by which the AFR can be regulated to targets throughout the engine's range of operation. This range was previously described in terms of operating parameters and engine parameters. From a controls perspective, the problem can be thought of as how to react to disturbances (i.e. changes in these parameters) that cause the AFR to err from its target. For example, a change in air throttle position (the initiation of a transient) causes a change in the manifold charge pressure (a disturbance), and hence the next cycle's AFR. This disturbance can occur very quickly. An example of a slow disturbance is the gradual rise in engine temperature following a cold start.

If the transient initiating the disturbances or the disturbances themselves can be measured, and the effect that the transient or disturbances have on the AFR is known, then the fuel pulse width can be modified to avoid an AFR error. This is feedforward control. Since feedforward uses information (ex. air throttle motion) that precedes the actual in-cylinder AFR error, fast, potentially zero error control can be achieved. To rely on feedforward alone requires the precise measurement of all parameters initiating transients or the disturbances that follow, a perfect understanding of their effect on the engine, and an exact knowledge of the characteristics of the actuators; a practical impossibility.

Feedback control directly measures the parameter to be controlled, in this case the AFR, and drives a modification of control variable(s), here the fuel pulse width, based on the magnitude of error. Feedback is fundamentally slower than feedforward because it has to wait for an error to occur. The advantage of feedback control is the ability to correct for unmeasured disturbances and system (i.e. engine, actuator) variations that arise out of changes in operating and engine parameters.

Figure 3 shows the topology of the lean-burn AFR control algorithm used in this study. Note that the air throttle position is not controlled by the ECS, however, its measurement (TVO_2) is used for control. The ECS regulates the AFR using the fuel pulse width (t_{inj}). The engine outputs of RPM, exhaust oxygen and cylinder pressure are sensed and used for control. Because to a large extent the success of a lean-burn engine control solution depends upon avoiding lean misfires, this structure was designed to provide fast, perhaps cycle-to-cycle, AFR control.

Figure 4 contains information as to where data was sampled, control calculations were performed, and outputs were set for AFR and MBT control. This figure shows that in contrast to all current production ECSs, sampling, calculations and outputs were performed with respect to crankshaft angle (not time). The advantage of this scheduling is that it tags data to a particular engine cycle which is necessary to provide cycle-to-cycle control.

Changes in air throttle position and engine speed cause fast disturbances that influence AFR. Feedforward, using measurements of TVO and RPM, was used to assist in avoiding AFR errors caused by these disturbances. This form of feedforward has been recently used on production vehicles (Bassi et al., 1985). For the test engine, manifold filling took less than one cycle. Manifold emptying could take more than one cycle depending on the severity of the transient. For engines with larger intake manifold volume-to-engine displacement ratios, modification of the feedforward would have to be made to account for the dynamics of manifold air filling and emptying.

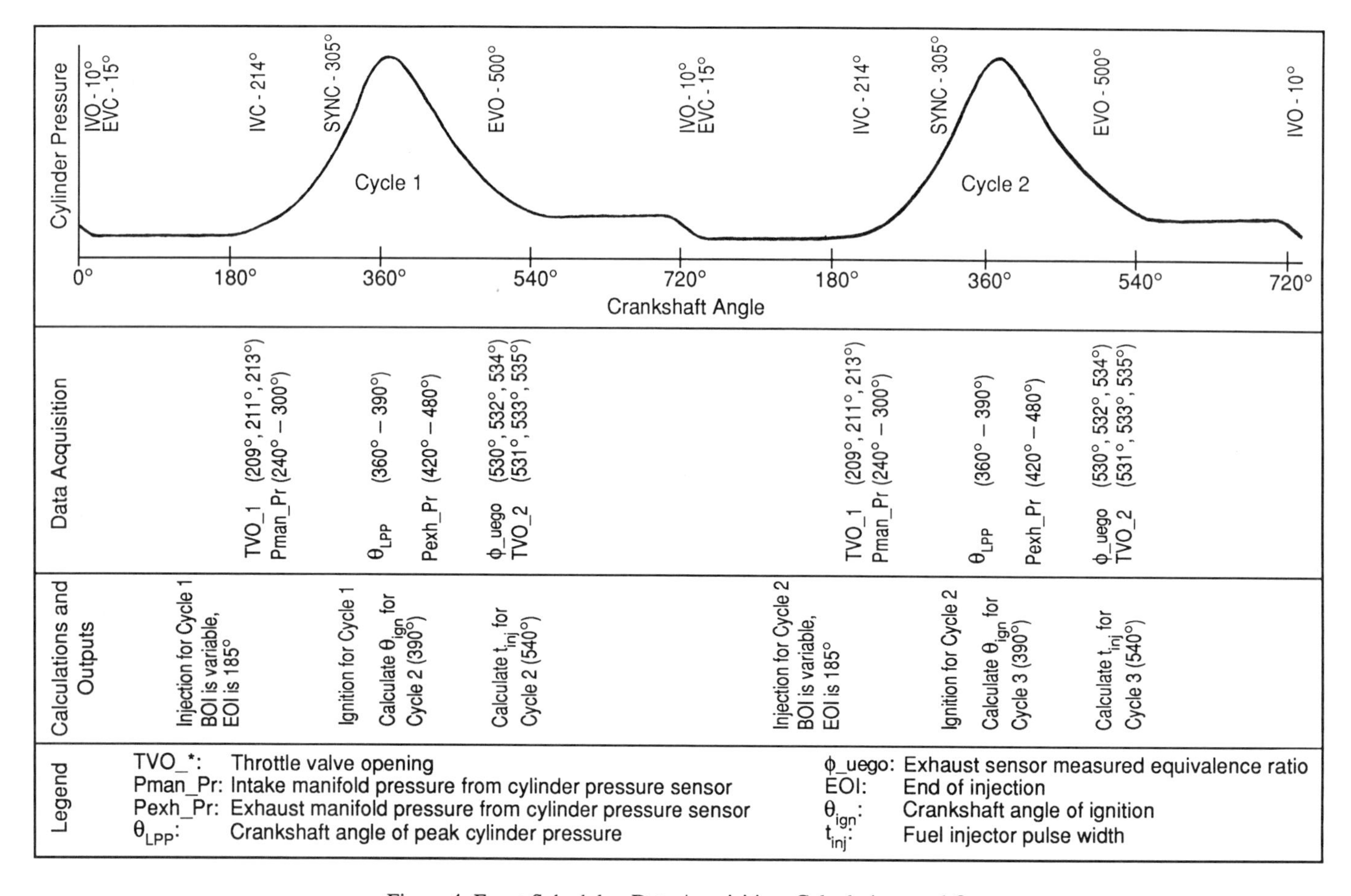

Figure 4: Event Schedule - Data Acquisition, Calculations and Outputs

Since other disturbances or system variations influencing AFR are either unmeasurable (ex. EGR ratio) or slow (ex. engine temperature), feedback was used to correct for their effect on AFR. Two feedback loops were tested. One of the loops used a UEGO (universal exhaust gas oxygen) sensor to measure the percentage of oxygen in the exhaust. For a known fuel type, a relationship between the exhaust's percentage oxygen and lean AFRs can be calculated. The UEGO sensor's characteristics and its use are well developed (Murase et al. 1988, Soejima and Mase 1985, Ishii et al. 1988, Tanaka et al. 1989). A sensor's response time limits the bandwidth of any feedback loop. The UEGO sensor's response time (10%-90%), using gasoline as a fuel in a multi-cylinder engine has been reported as 100 ms (Murase et al., 1988). Others, using a propane burner test stand, have shown it to be 400 ms (Soejima and Mase, 1985). For our single-cylinder test engine using natural gas as a fuel, the sensor's response time was measured as 400 ms (3 sensors were tested). With any fuel or engine, the response time of the this type of UEGO sensor will probably be slower than current production zirconia switch-type sensors (measured response time of approximately 45 ms).

The second feedback loop used a correlation between cylinder pressure and AFR to determine a cycle-to-cycle estimation of in-cylinder AFR. Several forms of these correlations have been identified (Houpt and Andreadakis 1983, Matekunis 1984, Gilkey and Powell 1985). A computationally efficient technique (Matekunis 1984, Gassenfeit and Powell 1989) was chosen for testing. For this technique, cylinder pressure data at known crankshaft angles is used to calculate a ratio of the process-averaged cylinder pressure before and after combustion (known as the pressure ratio, P-ratio or Pr). These process-averaged cylinder pressures are called Pman_Pr and Pexh_Pr. Thus Pr= Pman_Pr/Pexh_Pr. The basic relationship between the P-ratio and AFR can be calculated, but sensitivities of this relationship to operating and engine parameters require that it be determined experimentally. Figure 5a shows this relationship for the test engine. This relationship had sensitivities to RPM, TVO and LPP (location of peak cylinder pressure) as shown in Figures 5b, 5c, and 5d. Figure 5d was used to correct the measured P-ratio for variations in LPP from the target location. Figures 5b and 5c were used to determine target values of P-ratio (Pr_comm in Figure 3) corresponding to target values of AFR. Other, particularly thermal, sensitivities of the P-ratio-AFR relationship exist (Patrick and Powell, 1990).

In contrast to the UEGO sensor, the P-ratio provides a cycle-to-cycle estimation of AFR that follows the engine by less than an engine cycle. Also, unlike the exhaust sensor loop, increases in engine speed do not reduce the loop's effectiveness. However, the P-ratio loop has a problem in that at equivalence ratios greater than approximately 1.05 (this value is dependent on RPM and TVO) the slope of the P-ratio's sensitivity to AFR decreases and then changes in polarity. This can lead to a control loop instability. Because of this characteristic of a P-ratio or other AFR estimator based on cylinder pressure alone, an AFR control loop must use feedforward if significant changes in AFR can occur in one engine cycle. Both the UEGO sensor and the P-ratio loops used integral control laws.

MBT Control Logic

Previous work (Hubbard et al., 1975) identified a correlation between the location of peak cylinder pressure (LPP) and MBT over an engine's range of operation. This correlation was found to be valid for our test engine and constituted the MBT control logic (with an integral control law) used to set the ignition timing. It should be

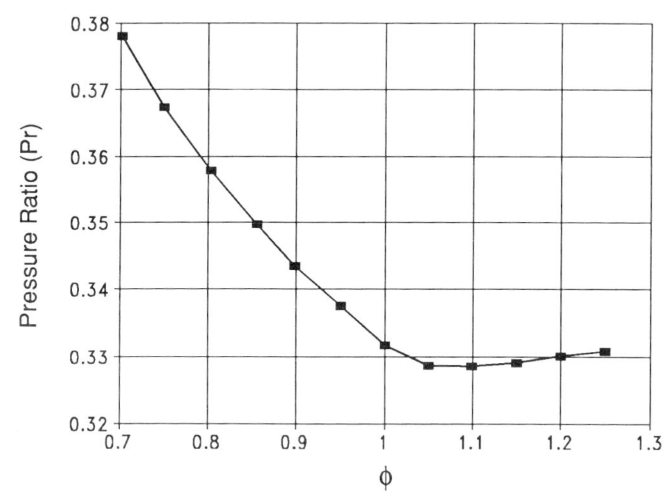

Figure 5a: Pressure Ratio vs Equivalence Ratio (1000 RPM, 40°TVO, 17.5°LPP, 1000 Cycle Average)

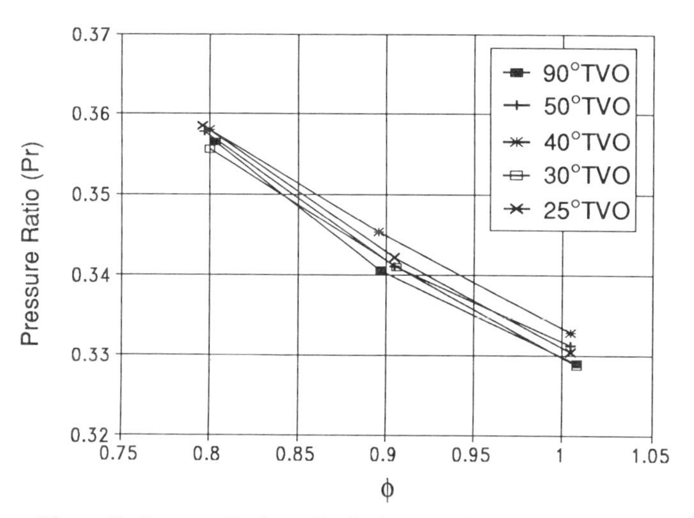

Figure 5b: Pressure Ratio vs Equivalence Ratio as a Function of Throttle Position (1000 RPM, 1000 Cycle Average)

noted that this correlation is essentially empirical and can be found not to be valid for particular engine/fuel combinations (Chang and Stevens, 1987). Because our test engine had a narrow range of optimum ignition timing (approximately 7°), feedforward was not used. Had its range been larger, feedforward using measurements of TVO and RPM would have been used to increase the engine's torque during rapid TVO and RPM transients.

RESULTS AND DISCUSSION

Figures 6 through 14 show the performance of various AFR control structures made from combinations of the feedforward and feedback control strategies shown in Figure 3. Most of these figures contain air throttle transients (at constant engine speed) because air throttle motion is the most common engine transient and has the fastest and greatest influence on AFR. For engines which operate near their lean limit, good AFR control during these transients is essential in order to avoid misfires.

For all tests, the engine was at operating temperature and the target AFR was ϕ=0.9. Although this AFR would not be lean enough for a production lean-burn engine, it was close to the lean limit for the test engine if EOI was not used as a control variable

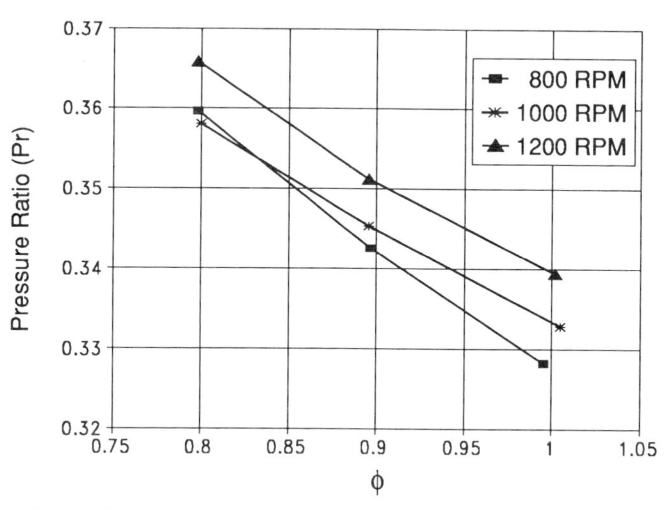

Figure 5c: Pressure Ratio vs Equivalence Ratio as a Function of Engine Speed (40°TVO, 1000 Cycle Average)

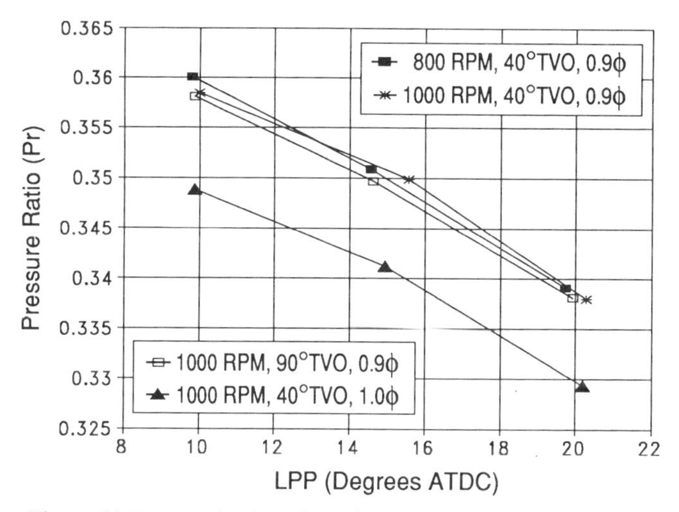

Figure 5d: Pressure Ratio vs Location of Peak Pressure at Various Engine Operating Points (1000 Cycle Average)

(which it was not). Therefore the salient problem of lean-burn engine control, that of keeping the engine from misfiring, exists for this test engine as it would for any other.

When examining the figures, the following must be kept in mind:

1. All data was taken, calculations were made, and outputs were set at specific crankshaft angles (see Figure 4).

2. At SYNC (305°) of each cycle, a data "snapshot" was taken of the latest values of sampled data, calculations and outputs from the previous period (SYNC to SYNC).

3. Intake manifold filling took less than one cycle while emptying took more than one cycle.

4. TVO_2 is the average of three air throttle position measurements made at 531°, 533°, and 535°. TVO_2 was used by feedforward to calculate the fuel pulse width. TVO_2's location was chosen because it was the latest crankshaft angle at which TVO

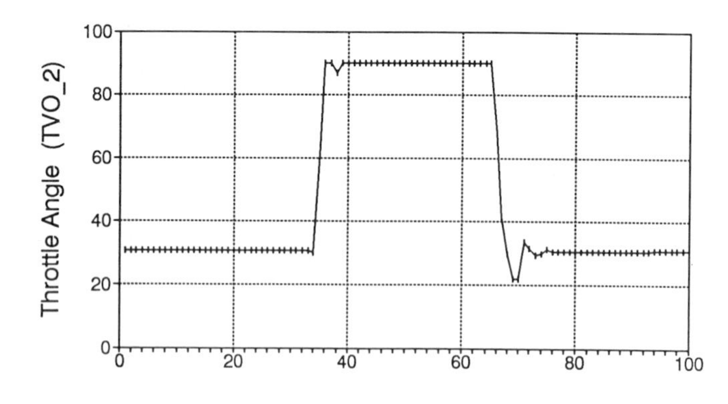

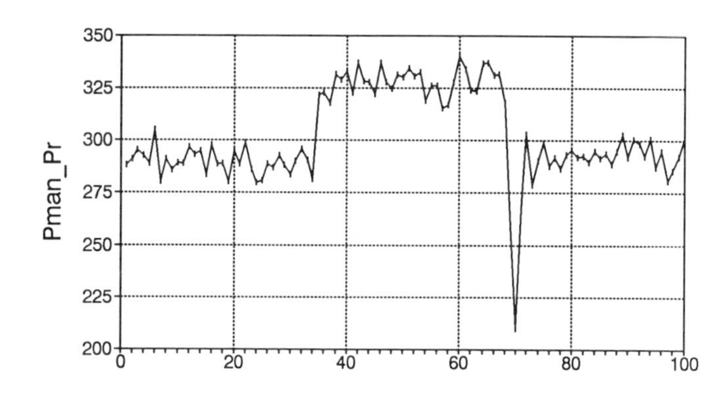

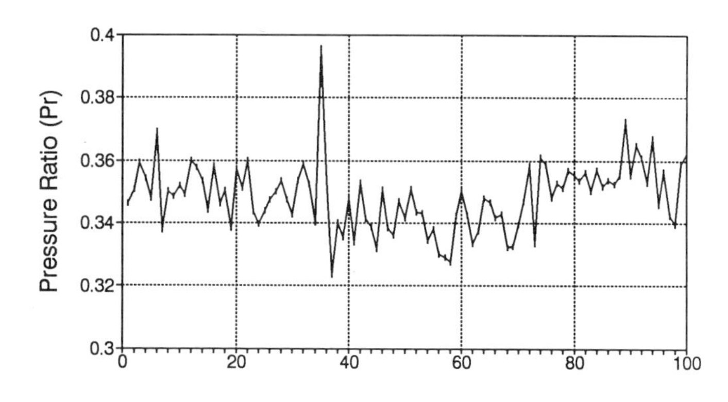

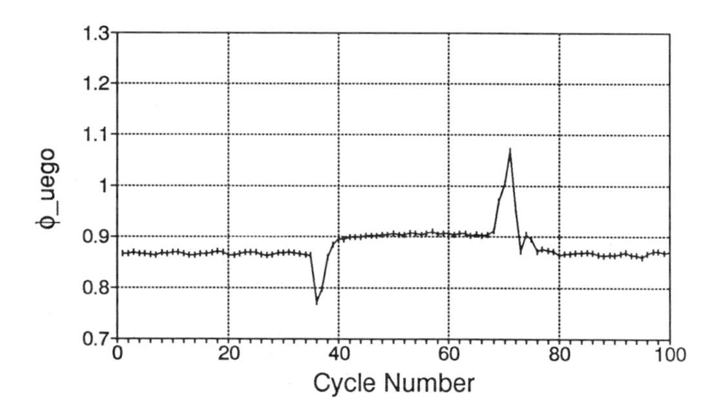

Figure 6: Air Throttle Opening and Closing Transients Using Feedforward Control (800 RPM)

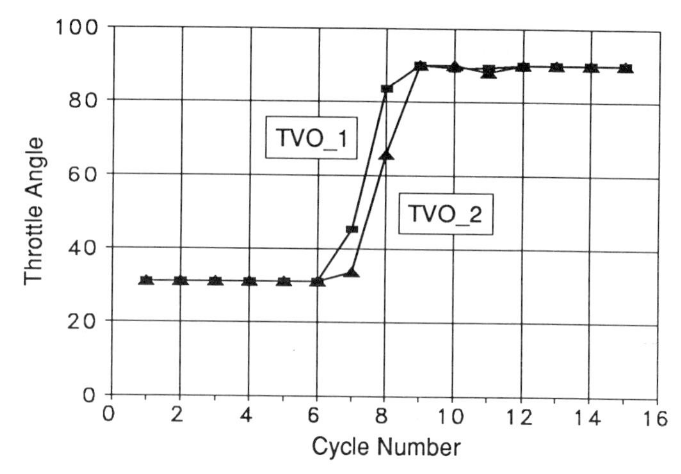

Figure 7a: Details of Throttle Opening Transient

information could be used for feedforward with a fixed EOI. In the feedforward algorithm, an increase in TVO_2 drove an increase in fuel pulse width to maintain the target AFR. However between 50° and 90°, the increase was very slight. TVO_2 ranged from 15° to 90° but the engine would not operate below approximately 22°.

5. Pman_Pr is the numerator of the P-ratio (Pr) and is proportional to the trapped mass in the cylinder.

6. Pr is proportional to the in-cylinder AFR. At AFR's leaner than stoichiometric, a Pr change of one graticule (0.02) implies that the AFR changed by approximately 13%. A higher Pr means a leaner mixture (unless ϕ is greater than approximately 1.05, see Figure 5a). A Pr of 0.4 indicates a misfire, since Pr was bound in software to be between 0.3 and 0.4. All Pr's shown in Figure 6 and Figures 8 through 14 were corrected for LPP not being in the target location (see Figure 5d).

7. ϕ_uego is the equivalence ratio determined from exhaust composition by the UEGO sensor. The response time of this sensor was approximately two cycles at these engine speeds.

Figure 6 shows TVO transients for which only feedforward control was used. Before and after the transients, the cycle-to-cycle AFR variations (according to the Pr) were on the order of 4% RMS. These variations were greater with this engine when gasoline was used as a fuel (approximately 5% RMS) and are on the order of those found by another researcher (Yu, 1963) using fast in-cylinder sampling valves (4% RMS with gasoline and a different engine). However, we do not believe that all of the cycle-to-cycle Pr variations are caused by AFR variations. At these engine speeds, the UEGO sensor was too slow to accurately follow cycle-to-cycle variations.

Despite careful engine mapping to determine the feedforward, a bias from the target (ϕ=0.9) existed at the low throttle condition. This bias could have resulted from an unmeasured disturbance (ex. injector pressure change) or a system variation (ex. TVO sensor bias). In any event, it shows that from a practical standpoint, AFR feedback is needed.

During this test, the UEGO sensor showed a lean excursion with throttle opening and a rich excursion with throttle closing. This was supported by the Pr output during throttle opening, but not during closing. The reason why the Pr did not show the rich transient is because the Pr's sensitivity at these AFRs is very low.

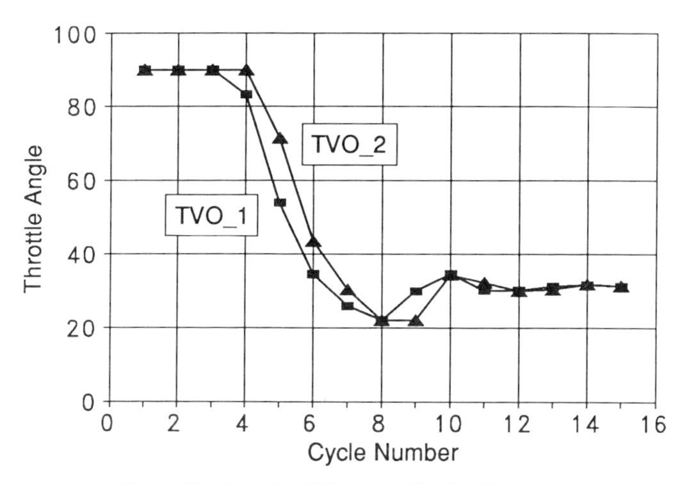

Figure 7b: Details of Throttle Closing Transient

Because Pr uses in-cylinder information, it indicated an AFR change before the UEGO sensor did. The AFR dynamics look like those from a liquid-fueled engine during the same transients, but are shorter in duration. According to the Pr trace, the AFR dynamics were contained within the period of air throttle opening. This is the key to understanding their cause.

Figure 7 shows TVO transients as a function of cycle number (from another test of this type). TVO_2 was used by the feedforward to calculate the fuel pulse width and TVO_1 was the throttle's position just before intake valve closing (IVC) (see Figure 4). TVO_2's location was chosen so as to allow time for computation and the injection duration. During throttle transients, the effective throttle position (in terms of actual trapped mass in the cylinder) for a given cycle was somewhere between TVO_2 and TVO_1. For the case of a throttle opening transient (Figure 7a), the effective throttle position would be greater than TVO_2. This means that a fuel pulse width calculated on the basis of TVO_2 would be too short and would result in a lean cycle. The smaller an engine's intake manifold volume-to-engine displacement ratio and the lower its speed, the more susceptible it would be to lean excursions during throttle valve opening transients. For the case of a throttle closing transient (Figure 7b), the effective throttle position would be less than TVO_2 resulting in a rich cycle. The smaller an engine's intake manifold volume-to-engine displacement ratio and the higher its speed, the more susceptible it would be to rich excursions during throttle closing transients.

Figure 8 shows TVO transients in which predictions as to the effective throttle position for each cycle were made from the trend of previous TVO_2s. The use of TVO_2 predictions reduced the AFR excursions. The difficulty in achieving these reductions in practice is in predicting the driver's input, particularly the initiation of the input, with the penalty of error possibly being another change in the input.

If multiple measurements of TVO were taken within a given cycle, and this information was used to drive a manifold and cylinder filling model, it would be possible to calculate the air entering the cylinder per cycle, regardless of throttle motion. Assuming this were possible (it is not at present), the problem remains how to modify injection events which are in progress, or what to do when events have just completed, while not falling prey to operational sensitivities such as those shown in Figures 1 and 2. For the case of an air throttle closing transient which begins after EOI and before IVC, the required action is to remove fuel already injected, which is impossible.

In practice, the simplest and most effective solution may be to schedule the air throttle dynamics by taking control of the throttle from the driver and assigning it to the engine control computer

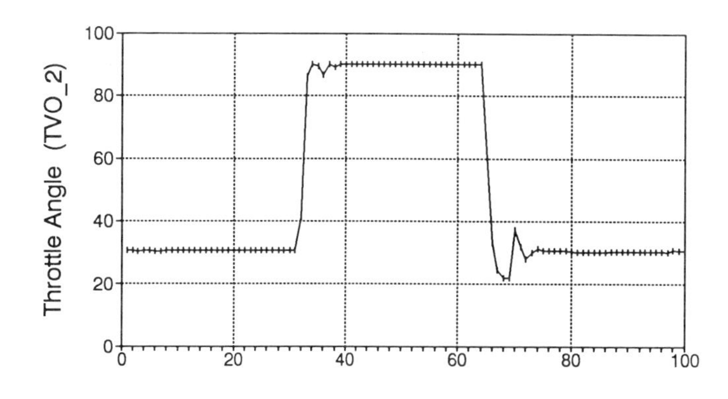

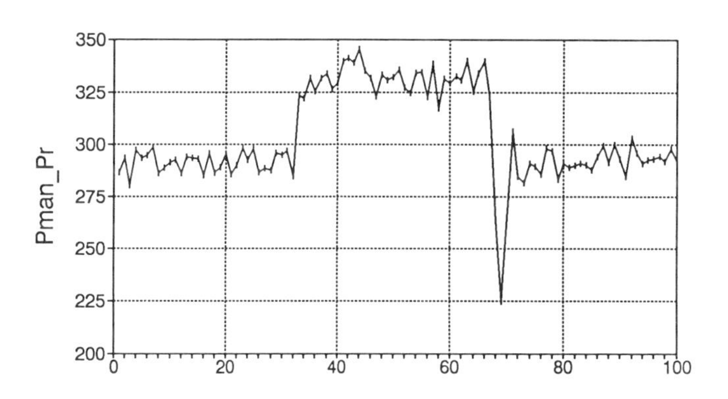

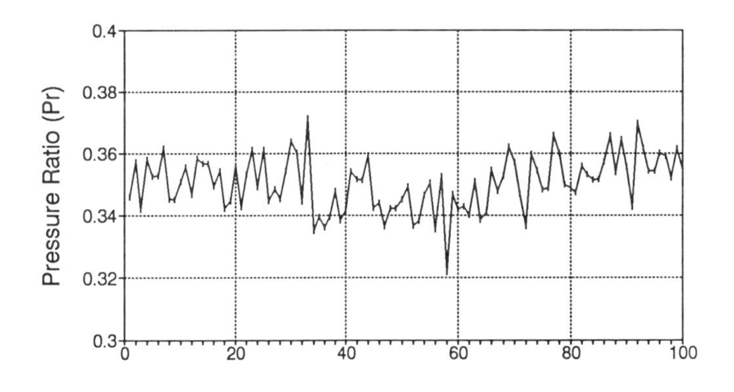

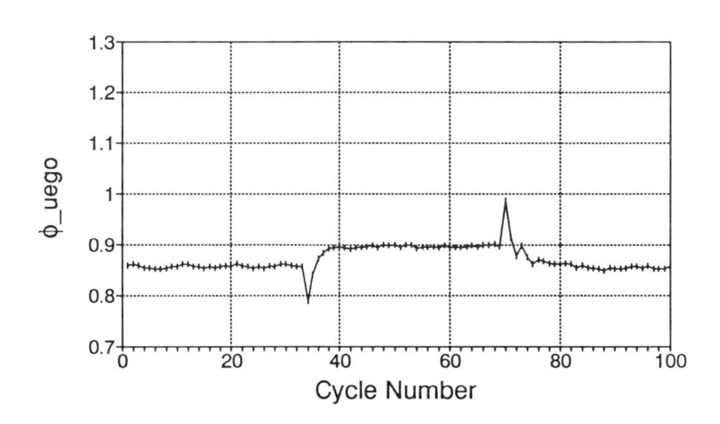

Figure 8: Air Throttle Opening and Closing Transients Using Feedforward with TVO Prediction (800 RPM)

(drive-by-wire). This way, air throttle motion could be gated to occur in coordination with piston motion, fuel pulse width, and fuel pulse phasing. In addition, electronic throttle control would allow the implementation of the second form of lean-burn ECS (see BACKGROUND).

Figure 9 shows TVO transients in which only the oxygen sensor feedback loop was active. This loop gave zero error at steady-state, but showed poor transient performance. The cycles in which the Pr reached the limit of 0.4 were misfires. The misfires could not be detected by the oxygen sensor because its response was too slow.

Figure 10 shows TVO transients with both feedforward and oxygen sensor feedback active. There were no steady-state errors but the settling time increased relative to the feedforward-only case. This is because the oxygen sensor loop had to "unwind" the correction it had added to the feedforward before the TVO transient occurred. Learning control would improve this system's performance.

Figure 11 shows the TVO transient performance of a feedforward and oxygen sensor feedback structure in which the feedforward was only driven by air throttle position. This considerably reduced the engine mapping required to determine the feedforward. The feedforward was "mapped" at 1000 RPM with the transient performed at 800 RPM. The performance of this configuration did not appreciably degrade as a result of this simplification. Figure 12 shows the performance of this structure during an engine speed transient. Note that 200 cycles of data are plotted. At the beginning of the transient, where the engine speed changed most rapidly, an AFR excursion was experienced, but following cycles show acceptable performance. At higher engine speeds, the oxygen sensor loop gain was too high resulting in some instability, but this could be removed through gain scheduling.

Figure 13 shows the TVO transient performance of the Pr feedback loop with no feedforward. At steady-state, there was a bias which was primarily the result of a cylinder pressure sensor voltage bias shift. This bias could be reduced by electronically grounding the sensor each engine cycle. The gain of this loop was high resulting in oscillations at steady-state. Even though this loop used cycle-to-cycle AFR information, as a feedback loop it needed to see an error before it could provide any correction. The throttle closing AFR recovery was slower than that for throttle opening because at throttle closing, the AFR became rich where the Pr's sensitivity to AFR is reduced. The Pr loop showed transient performance similar to that of the oxygen sensor (Figure 9). If the engine speed at which this transient occurred was increased however, there would be a steady degradation in the performance of the oxygen sensor loop but not the Pr loop.

In Figure 14, the Pr loop was augmented with feedforward. The addition of feedforward reduced the magnitude of AFR excursions and the AFR recovery time. However, the addition of the Pr loop did not give improvements in transient performance over the case of feedforward alone (Figure 6). Note also that fortuitously, the steady-state bias error in the feedforward cancelled with the bias of the Pr. In general, the Pr loop would not give zero steady state AFR error because of inaccuracies in our knowledge of the Pr-AFR relationship.

CONCLUSIONS

The conclusions of this work are summarized as follows:

1. A timed fuel injection system can allow leaner operation of a natural gas fueled SI engine than a continuous flow system. For the engine tested, the use of a timed fuel injection system permitted an 8% increase in the lean limit AFR.

2. Increases in injection pressure for a timed fuel injection system can allow leaner engine operation.

3. The use of phasing of the fuel pulse width as a control variable can extend an engine's lean limit of operation. For the engine tested, at certain operating conditions, a change in EOI of 20° moved the engine from a best stability point to a no-run condition.

4. The location of peak pressure (LPP) method of ignition timing can provide MBT for natural gas operation with AFRs lean of stoichiometric. For the engine tested, a 17.5° LPP target was best with natural gas at ϕ=0.9. Previous work has shown that a 15° target is best with gasoline at ϕ=1.0 (Hubbard et al., 1975).

5. With feedforward AFR control, air throttle motions can cause AFR errors by means of their modification of the air flowrate into the cylinder between when fuel pulse width calculations are made and IVC.

6. If AFR errors caused by air throttle motion are severe, or if EOI is to be used as a control variable, then electronic throttle control may be the easiest way to avoid AFR errors or misfires during load transients. This actuator may provide the key to cycle-to-cycle AFR control.

7. The summation of integrated error from AFR feedback with feedforward will reduce steady-state error to zero but can increase AFR excursions during transients. To avoid this problem, corrections from AFR feedback should be used to modify the feedforward itself (i.e. learning control).

8. Cylinder pressure based AFR feedback can provide fast AFR control. If this technique uses pressure ratios, then mapping must be performed to determine the Pr-AFR relationship and its sensitivities. Uncompensated sensitivities result in steady-state AFR errors. The reduction in sensitivity and subsequent polarity reversal of the Pr-AFR relationship at rich AFRs requires that this loop be used in conjunction with TVO-driven feedforward to avoid transitions to rich AFRs.

9. With further engineering development, current production-like sensors and actuators (for cylinder pressure, see Kondo et al. 1975, Antastasia and Pestana 1987) could be used to construct lean-burn natural gas ECSs, and make the lean-burn natural gas engine solution a reality.

ACKNOWLEDGEMENTS

The authors gratefully acknowledge the support of this work by Nissan Motor Co. Ltd.. Thanks also to Nick Fekete who assisted in the design and fabrication of the induction and fuel systems.

REFERENCES

Antastasia, C.M. and Pestana G.W., 1987, "A Cylinder Pressure Sensor for Closed Loop Engine Control", SAE Paper No. 870288.

Bassi, A. et al., 1985, "C.E.M.-The Alfa Romeo Engine Management System-Design Concepts-Trends for the Future", SAE Paper No. 850290.

Chang, M.-F. and Stevens, J.E., 1987, "Location of Peak Pressure for an Axially Stratified-Charge Engine", SAE Paper No. 870080.

Gassenfeit, E.H. and Powell, J.D., 1989, "Algorithms for Air-Fuel Ratio Estimation Using Internal Combustion Engine Cylinder Pressure", SAE Paper No. 890300.

Gilkey, J.C. and Powell, J.D., 1985, "Fuel-Air Ratio Estimation from Cylinder Pressure Time Histories", *J. of Dynamic Systems, Measurement and Control*, Dec. 1985.

Houpt, P.K. and Andreadakis, S.K., 1983, "Estimation of Fuel-Air Ratio from Cylinder Pressure in Spark Ignition Engines", SAE Paper No. 830418.

Hubbard, M. et al., 1975, "Closed Loop Control of Spark Advance Using a Cylinder Pressure Sensor", ASME Paper 75-WA/Aut-17.

Ishii, J. et al., 1988, "Wide Range Air-Fuel Ratio Control", SAE Paper No. 880134.

Kondo, M. et al., 1975, "Indiscope-A New Combustion Pressure Indicator with Washer Transducers", SAE Paper No. 750883.

Matekunis, F.A., 1984, "Engine Combustion Control with Ignition Timing by Cylinder Pressure Management", U.S. Patent No. 4,622,939, Nov. 14, 1984.

Murase, I. et al., 1988, "A Portable Fast Response Air-Fuel Ratio Meter Using an Extended Range Oxygen Sensor", SAE Paper 880559.

Patrick, R.S. and Powell, J.D., 1990, "A Technique for the Real-Time Estimation of Air-Fuel Ratio Using Molecular Weight Ratios", SAE Paper No. 900260.

Soejima, S., Mase, S., 1985, "Multi-Layered Oxygen Sensor for Lean Burn Engine Application", SAE Paper No. 850378.

Stivender, D.L., 1978, "Engine Air Control-Basis of a Vehicular Systems Control Hierarchy", SAE Paper No. 780346.

Tanaka, H. et al., 1989, "Wide-Range Air-Fuel Ratio Sensor", SAE Paper No. 890299.

Yu, H.T.C., 1963, "Fuel Distribution Studies: A New Look at an Old Problem", SAE Transactions, Vol. 71, pp 596-613.

APPENDIX A: DESCRIPTION OF APPARATUS

Engine

- single cylinder CFR
- 611.7 cm^3 (37.33 $in.^3$) displacement
- 8.26 cm (3.25 in.) bore, 11.43 cm (4.5 in.) stroke
- 6.0:1 compression ratio
- 275 to 2200 RPM range
- single spark plug
- single shrouded intake valve
- single exhaust valve
- intake system:
  - 24.23 cm x 3.18 cm dia. (9.54 x 1.25 in. dia.) pipe to throttle
  - volume 40% of swept cylinder volume
- exhaust system:
  - 2.44 m x 5.08 cm (8 ft. x 2 in.) diameter pipe
  - empties into unrestricted volume

Fuel

- methane (commercial grade):
  - 93% CH_4
  - balance H_2O, N_2, O_2, light hydrocarbons

Sensors

1. Throttle position (resistive type)
2. Cylinder pressure (laboratory grade)
3. Universal Exhaust Gas Oxygen Sensor (UEGO):
   - measurement range: $0.4 < \phi < 2.0$
   - accuracy 0.7 %
   - 10%-90% response time 400 ms (experimentally determined)
4. Crankshaft encoder (one degree resolution)
5. Camshaft encoder (one pulse per engine cycle)

Actuators

1. Fuel injector:
   - General Motors production "TBI" type used with 2.5 liter gasoline engine
   - 2.8 scfm @ 80 psig upstream pressure
   - injector mounted in aluminum block
   - injector block located near intake port or near air throttle
2. Ignition system (capacitance discharge)

Engine Control Computer

In-house design based on Intel 16-bit microcontroller

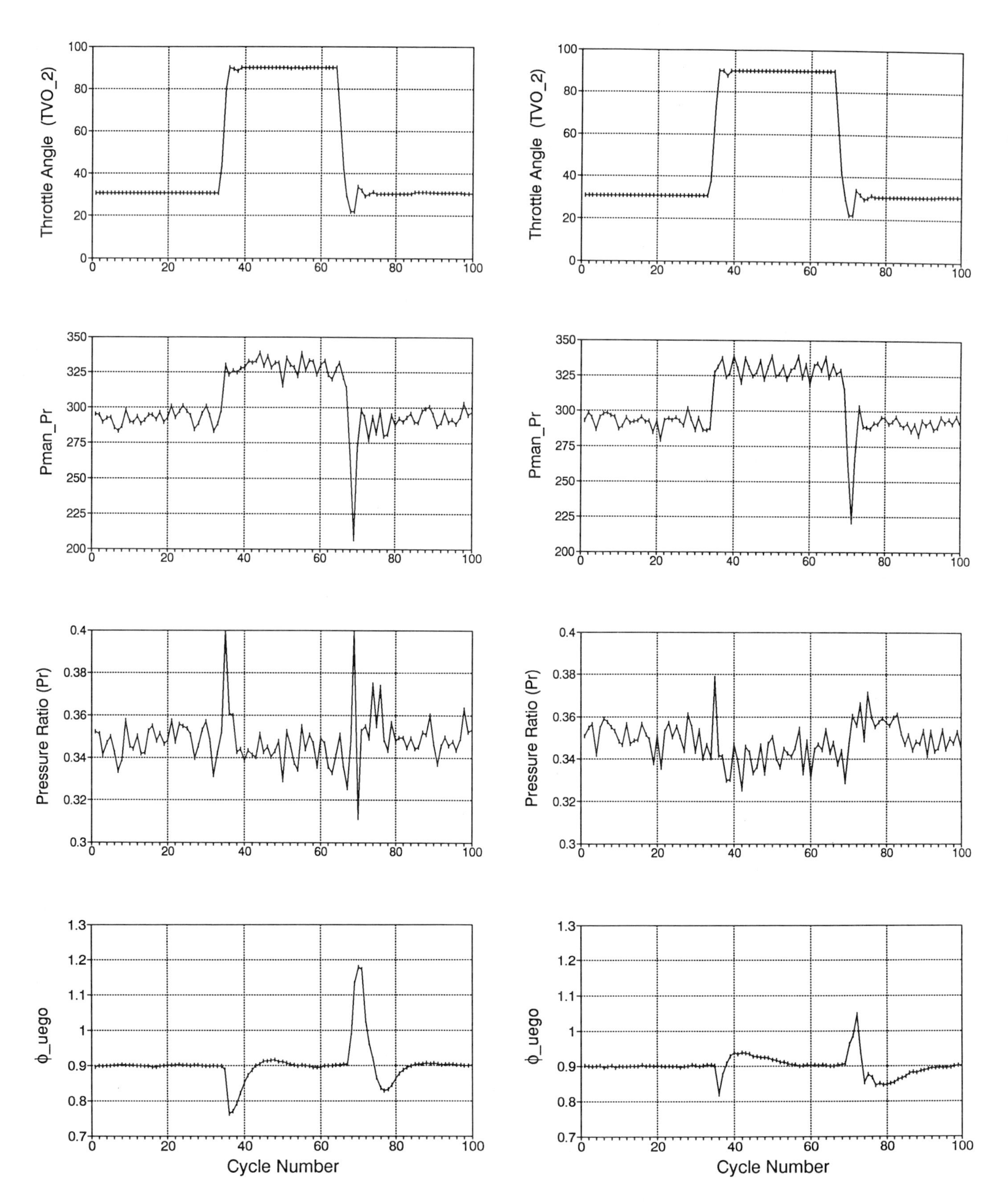

Figure 9: Air Throttle Opening and Closing Transients Using Exhaust Oxygen Sensor Feedback (800 RPM)

Figure 10: Air Throttle Opening and Closing Transients Using Feedforward and Exhaust Oxygen Sensor Feedback (800 RPM)

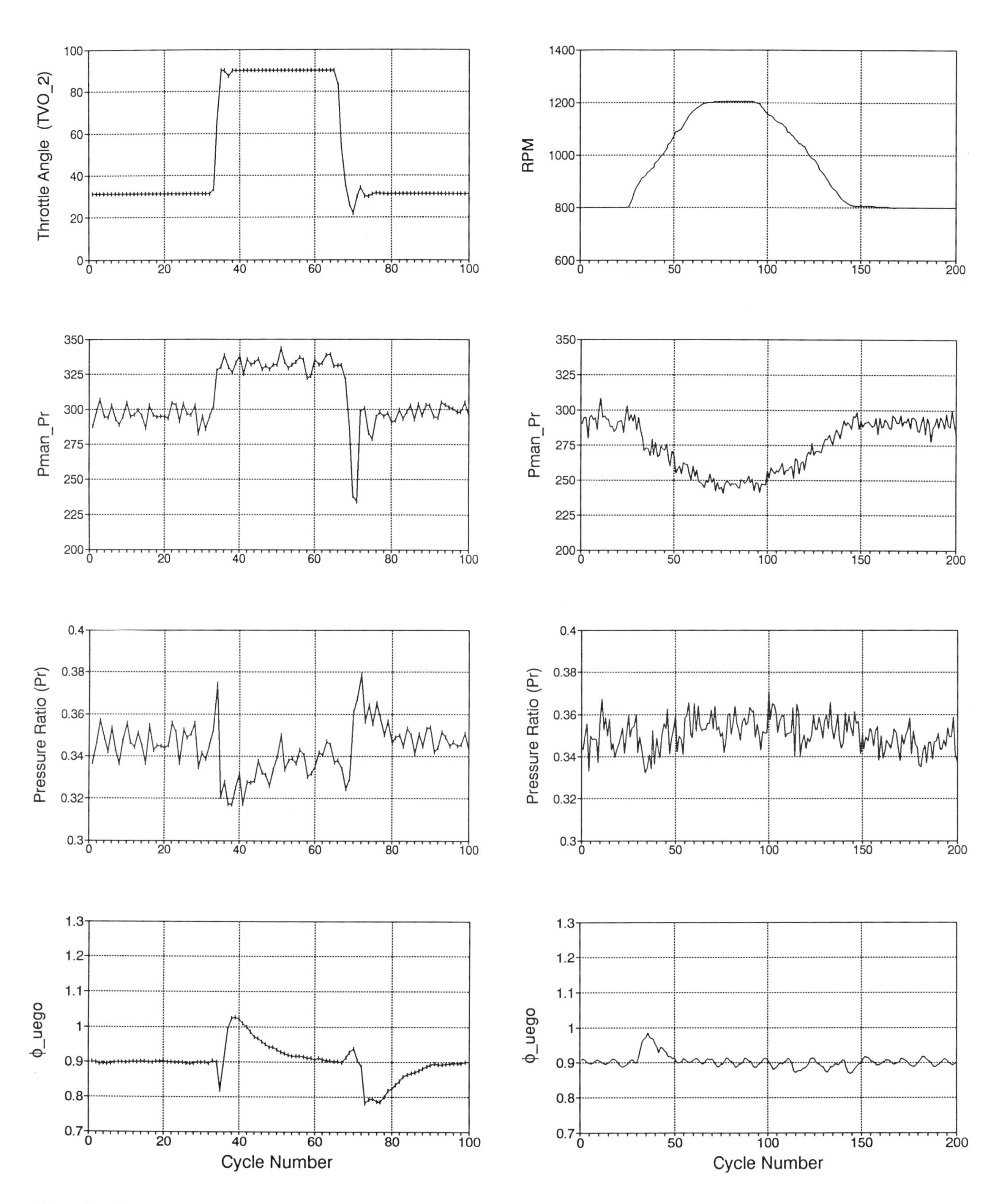

Figure 11: Air Throttle Opening and Closing Transients Using TVO-Only Driven Feedforward and Exhaust Oxygen Sensor Feedback (800 RPM)

Figure 12: RPM Transient Using TVO-Only Driven Feedforward and Exhaust Oxygen Sensor Feedback (30° TVO)

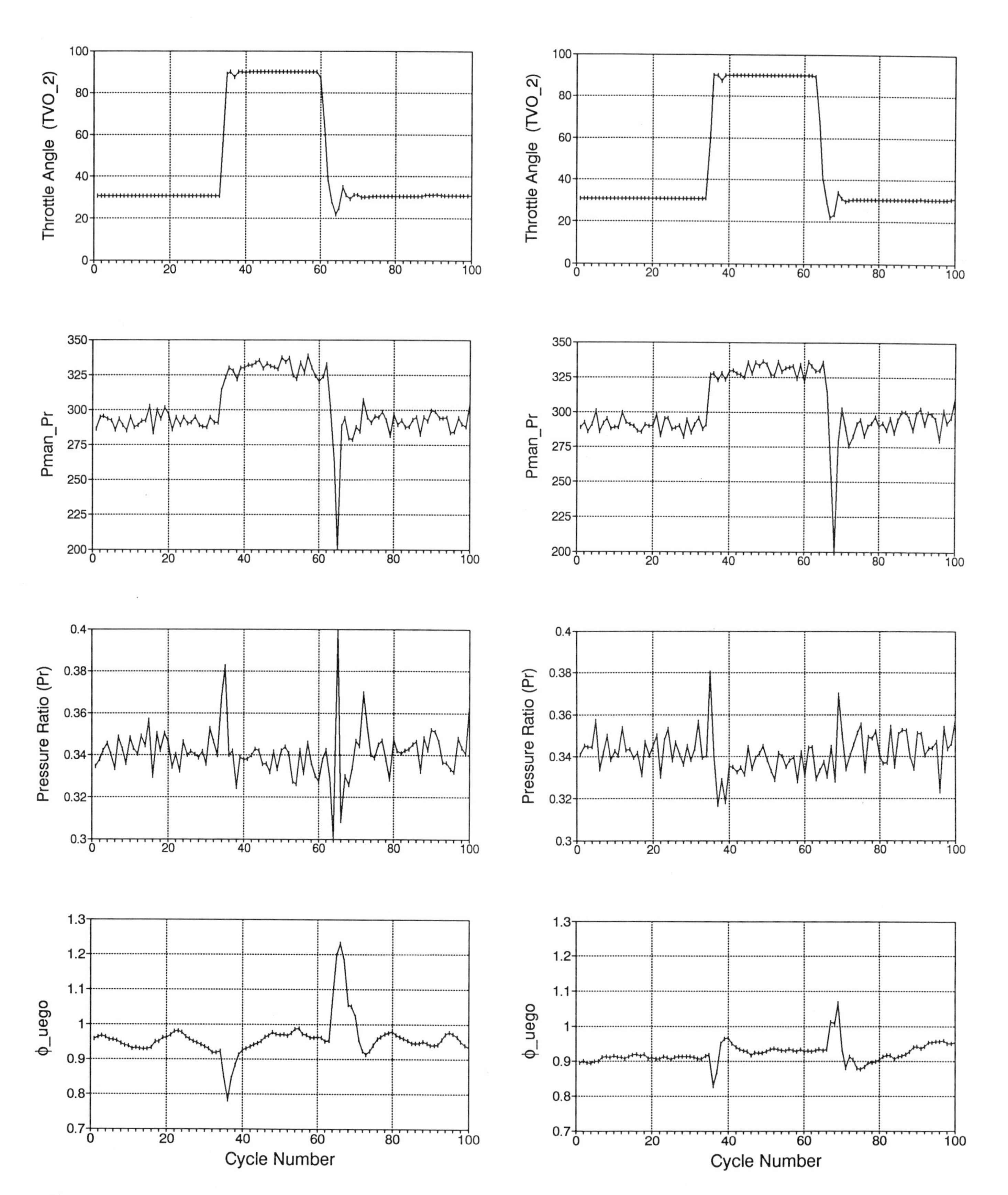

Figure 13: Air Throttle Opening and Closing Transients Using Pressure Ratio Feedback (800 RPM)

Figure 14: Air Throttle Opening and Closing Transients Using Feedforward and Pressure Ratio Feedback (800 RPM)

ICE-Vol. 15, Fuels, Controls, and Aftertreatment
For Low Emissions Engines
ASME 1991

LOW COST SENSOR FOR ELECTRONIC FUEL METERING IN SMALL ENGINE APPLICATIONS

Jeffery C. Beall and Robert G. Leonard
Department of Mechanical Engineering
Virginia Polytechnic Institute and State University
Blacksburg, Virginia

ABSTRACT

Electronic fuel metering has greatly reduced emissions and improved fuel economy in automotive applications. A central issue in the transfer of this technology to millions of one and two cylinder utility engines is the high cost of sensors and actuators. This paper describes methods by which engine speed, gauge manifold pressure and manifold pressure transients can be derived from intake manifold pressure fluctuations using a piezo-film type microphone element.

This very low cost transducer, coupled with a microprocessor, provides a practical means for electronic fuel metering in small engine applications. Microprocessor control offers the flexibility to use alternate fuels and to reduce emissions through better control of A/F ratio.

NOMENCLATURE

P_a	atmospheric pressure
P_m	intake manifold pressure
$p(t)$	acoustic pressure signal
R	diaphragm radius
S_o	sensitivity (displacement/pressure)
t_d	diaphragm thickness
η_o	diaphragm displacement at center
ν_d	circumferential stress
ν_o	initial circumferential stress

INTRODUCTION

In the application of electronic fuel metering to small engines, economic considerations have a significant impact on the design of the hardware. It is desirable to use the minimum number of sensors and to utilize low cost sensors to determine the information necessary to meter the fuel. The primary information needed is the rate of mass air flow into the engine.

Mass air flow can be measured directly using a hot wire or hot film thermal mass sensor as found in Bosch systems (Felger and Plapp, 1985), or estimated indirectly from one of the methods below.

Venturi effect - as found in carburetors, fuel metered as a function of air volume flow. Rarely compensated for temperature.

Vane type flow meter - measures volume air flow. Requires compensation for temperature and atmospheric pressure.

Throttle-Speed method - requires a calibration map of engine air flow as a function of engine speed and throttle angle. Must be compensated for temperature and changes in atmospheric pressure.

Speed-Density method - requires a calibration map of engine air flow as a function of speed and manifold absolute pressure. Must be compensated for temperature.

The speed-density method has been widely used because it is relatively easy to make reliable measurements of engine speed, air temperature and manifold pressure. Figure 1 shows an engine calibration map for the test engine using this method. (Engine specifications are included in the figure.) The map is accurate for a specific engine under steady conditions at the calibration temperature. In normal use, absolute manifold pressure and engine speed (using an interpolation if needed) are used to extract the nominal mass air flow from the map. This value is adjusted for the current manifold air temperature and then used to calculate fuel rate.

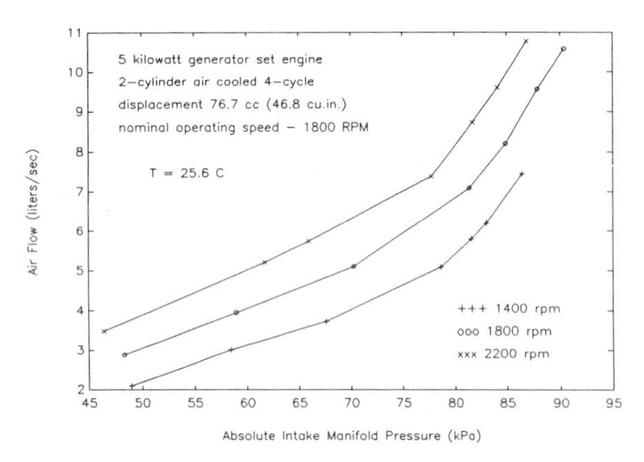

Figure 1 Engine Calibration Map - Air Flow vs. Manifold Pressure at 1400, 1800 2200 RPM.

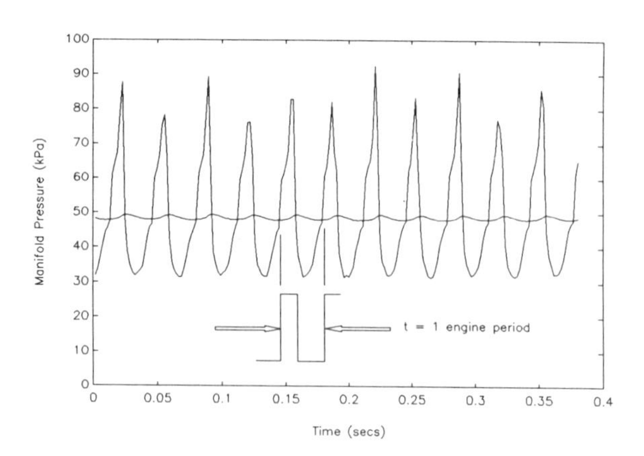

Figure 2 Estimate of engine period from manifold pressure signals.

As stated above, the engine calibration map is for steady conditions. Mass air flow differs from the nominal map value during throttle transients. For example, when the throttle is opened suddenly, extra air flow is required to pressurize the manifold volume to its new steady value. Moreover, the rapid increase in manifold pressure reduces fuel vaporization rate. These two effects produce a lean mixture during acceleration unless compensation is provided. These transient conditions can be identified and compensated for by estimating the rate of change of manifold pressure.

In addition, for engines with one or two cylinders, engine speed can be derived from the manifold pressure signal. Figure 2 shows a plot of the pressure waveform as found in the test engine after low pass filtering. The periodic pressure fluctuations are caused by the intake stroke which occurs once during each engine revolution (once every other revolution for a 1-cylinder engine). The apparent differences in adjacent cycles result from placement of the pressure sensor closer to one of the cylinders. The waveform can be averaged to generate its mean value (also shown). The signals can then be differenced to produce a square wave signal with a period equal to the inverse of the engine speed in revs/sec.

The discussion above indicates that the speed-density method can provide steady mass air flow data and transient compensation if average absolute manifold pressure, rate of change of manifold pressure, engine speed, and manifold temperature are measured. In small engine applications all of this information (except air temperature) is contained in the absolute manifold pressure waveforms.

The purpose of this paper is to show that the required information can also be extracted from the engine manifold pressure using a very inexpensive microphone element - with one difference. The pressure extracted using a microphone is the manifold gauge pressure (not absolute) and therefore the engine calibration map has both a reference temperature and pressure. If it is necessary to provide compensation for changes in engine operation resulting from changes in atmospheric pressure (e.g. high altitudes) the map can be scaled for local conditions.

MICROPHONE FUNDAMENTALS

The microphone elements of interest in this work (i.e. low cost) include the condenser or capacitor microphone, the electret microphone and the polyvinylidene fluoride (PVDF) film type microphone. Any of the microphone types can be used - however the PVDF film microphone is preferred because of its simplicity, low cost and the high strength of its diaphragm. The fundamental operation of each type is reviewed below.

The condenser (or capacitor) microphone consists of two parallel plates separated by a dielectric (usually air). A large external polarizing voltage is applied across the plates resulting in a charge stored on the plates. The back plate is rigid and stationary while the front plate is a thin metallized plastic diaphragm. Acoustic pressure fluctuations striking the front plate cause it to move relative to the stationary plate. The resulting change in plate separation (capacitance) produces a change in voltage proportional to the pressure fluctuations. An internal amplifier is often included in the microphone since the signal is too weak to be transmitted any distance (Eargle,1981).

The electret microphone operates as above but uses an electret material for the diaphragm. Electret materials hold a charge without the need for a large external polarizing voltage. Electret microphones have replaced condenser microphones for use in most practical applications (Eargle,1981).

PVDF film transducers operate on a different principle. PVDF is a semi-crystalline polymer. The beta-phase of the film is metalized and polarized (using a high gradient electric field) to enhance its piezoelectric behavior. A mechanical strain of the material produces a voltage on the metalized surfaces. To construct a microphone element, the film is stretched across a ring to form a diaphragm. Pressure waves stress the film, producing a voltage which is amplified by an FET amplifier built into the microphone (Wang et al,1988).

The typical bandwidth of the microphones described above ranges from 20 Hz to at least 10 kHz. Since manifold pressure transients are between 20 and 200 Hz, changes in microphone characteristics over frequency are not significant (see also Figure 10).

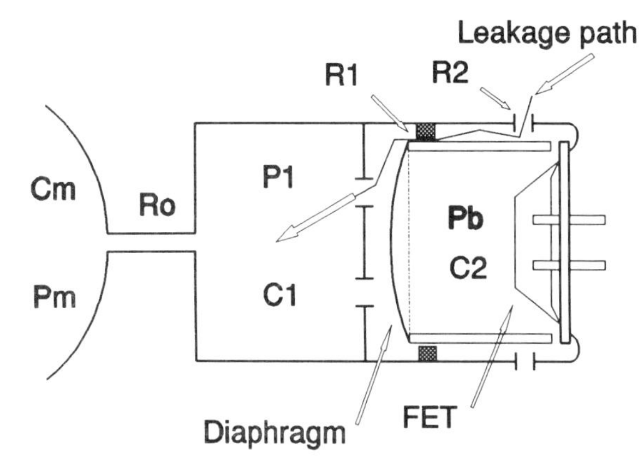

Figure 3 Microphone Element attached to the Intake manifold through orifice Ro.

MEASURING MANIFOLD PRESSURE

The microphones described above transduce extremely small changes in pressure. A typical minimum sensitivity is 1 millivolt per microbar. In contrast the typical pressure variations in the intake manifold range from 7-55 kPa (1 kPa=10000 microbars). The microphone is attached to the manifold as shown in Figure 3. The orifice (Ro) and front cavity (C1) act as a filter to attenuate the large engine pressure fluctuations and prevent saturation of the microphone output. The average front cavity pressure (P1) is equal to the average manifold pressure (Pm). The backcavity pressure (Pb) depends on the two leakage path resistances (R1 and R2), the manifold pressure (Pm) and atmospheric pressure (Pa).

The manifold/microphone system can be represented by the analogous resistor and capacitor network shown in Figure 4, where pressure corresponds to voltage, air flow to current and flow resistance to electrical resistance.

When the housing (at R2) is sealed, the backcavity pressure will equalize with the static front cavity pressure via leakage through R1. This is desirable in normal use - to prevent gross pressure changes (wind gusts, etc.) from displacing the diaphragm. Microphone packages are often vented to insure this equalization. When connected to the intake manifold as shown in Figure 3, with the housing sealed, the backcavity pressure equals

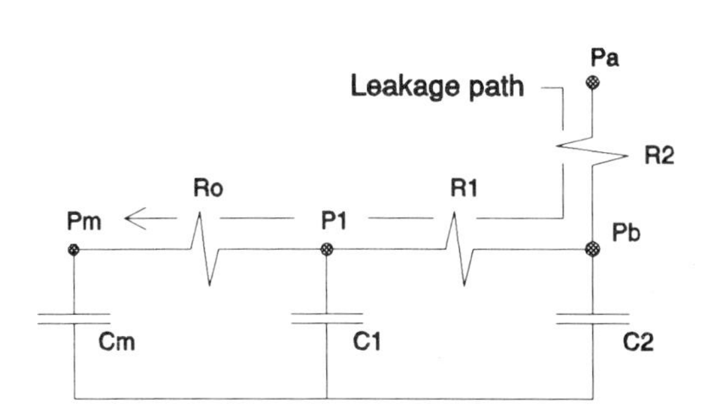

Figure 4. Equivalent RC-Network for the microphone and manifold.

the front cavity pressure (Pm).

When the housing is not completely sealed, air will be drawn through R2 and R1 by the low pressure in the front cavity. As a result, the backcavity pressure will vary depending on R1 and R2. This can be seen as a voltage divider network in Figure 4. If R1<<R2, the backcavity pressure will approach P1. If R2<<R1, the backcavity pressure will be nearer Pa. In this case, the diaphragm will be stretched by the difference in pressure, Pb-Pm (see also Figure 3).

PRINCIPLE OF OPERATION

Housing Sealed

The original intent of this investigation was to determine if the peak-to-peak (p-p) acoustic signal from the intake manifold could provide information about air flow. Specifically, Figure 5 shows plots of the manifold absolute pressure fluctuations for the test engine. Six steady-state generator load conditions are shown, from no-load through full-load at constant speed (1800 rpm). Full-load on the generator corresponds to about a 90% load on the engine. The pressures shown were measured with an absolute pressure sensor.

This graph shows immediately that the p-p pressure fluctuations change very little from one operating point to the next. In addition the change in p-p amplitude is not monotonic.

This is shown at the bottom of figure 5, where the standard deviation of the raw pressure signal has been calculated and plotted to produce a measure of the p-p amplitude. As shown, it is not possible to distinguish between operation near no-load and operation near full-load from the p-p amplitude.

With the microphone housing sealed, it is unlikely that reliable information about air flow can be determined from the p-p pressure fluctuations in the manifold.

Housing Not Sealed

Measurements made with the housing not

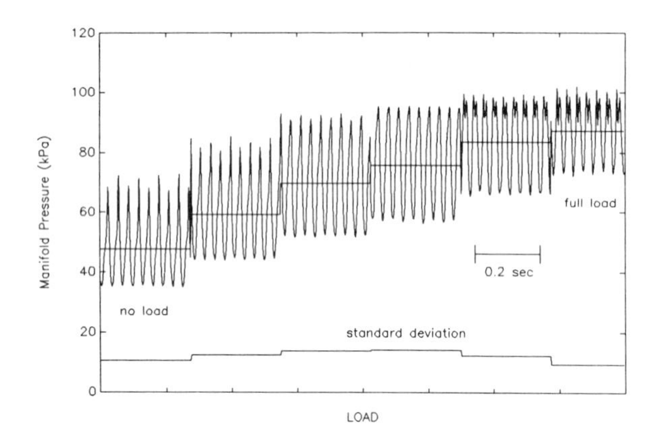

Figure 5 Peak to peak manifold pressure with mean value and standard deviation from the mean for six load cases.

sealed produce a more interesting result. In this case the diaphragm is strained by the difference in pressure Pb-Pm (or Pa-Pm, since Pb is nearly equal to Pa). At engine no-load conditions the average pressure difference is large (Pa-Pm = 47 kPa (14 in Hg)) and the diaphragm has a large net displacement. Under full-load, the manifold pressure is very nearly atmospheric (Pa-Pm = 3-6 kPa (1-2 in Hg)), and the net diaphragm displacement is small.

The net displacement of the diaphragm changes the sensitivity of the microphone. A large pressure difference stretches the membrane tighter (increases the circumferential stress) and reduces the sensitivity to the acoustic pressure signal, p(t). The displacement of the microphone diaphragm is a function of the circumferential stress as given in (1) below, adapted from Merhaut (1981).

$$\eta_o(t) = \frac{R^2 p(t)}{4\nu_o} \tag{1}$$

Circumferential stress for a thin membrane of this type can be calculated using the elementary equation (2) below, adapted from Cook and Young (1981).

$$\nu_d = \frac{R(P_a - P_m)}{2t_d} \tag{2}$$

By substitution of (2) into (1) (i.e. $4\nu_o \rightarrow 4(\nu_o + \nu_d)$) we can see that microphone sensitivity is inversely proportional to Pa-Pm.

$$S_o = \frac{\eta_o(t)}{p(t)} = \frac{R^2}{4\nu_o + \frac{2R(P_a - P_m)}{t_d}} \tag{3}$$

Figure 6, shows both the actual increase in microphone sensitivity as Pm gets larger (i.e. Pa-Pm gets smaller) and the ideal change in sensitivity as calculated from equation (3). In using equation (3), $4\nu_o$ was assumed negligible compared to the other terms and a normalizing factor was used to match the curves at Pm = 48.4 kPa.

In normal microphone operation the diaphragm is never subjected to the differential pressures in use here. However, stress calculations (using (2)), with a differential pressure of 104 kPa, indicate a maximum stress of 10 MPa. The minimum yield strength is 20 MPa, providing a safety factor of 2. (Film yield strength is rate, temperature and axis dependent, but 20 MPa is considered conservative by the manufacturer for this application.)

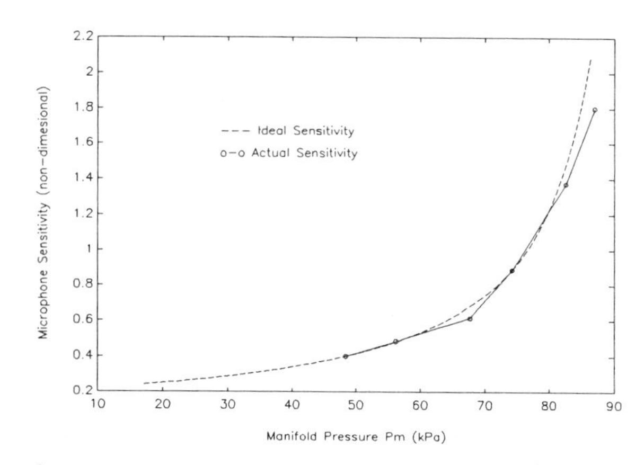

Figure 6 Microphone sensitivity as a function of manifold pressure.

Figure 7 shows the p-p voltage output of the microphone as a function of the exciting pressure. The microphone output is well behaved for the lower three loads. However, as the periodic displacements become larger, the voltage output from the microphone changes dramatically. This effect can be attributed to properties of the PVDF film.

The film has a linear relationship between strain and voltage output for very small strains (displacements). As the strain increases, voltage and strain are related by a set of hysteresis curves. Figure 8 contains a phase plot of voltage vs. strain (actually voltage vs. pressure) for the microphone at each of the load points. Even at the lowest load point (near 50 kPa in Fig.8), the strain vs. voltage plot is not linear, but nearly circular (indicating about 90 degrees of phase shift). For higher loads, the hysteresis properties become more pronounced.

MICROPHONE OUTPUT

The sensitivity of the microphone changes with differential pressure Pa-Pm. It is this change in sensitivity which allows the microphone to be used as a sensor for

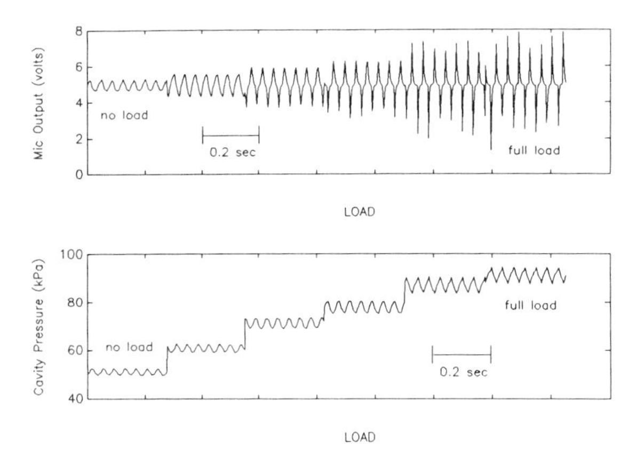

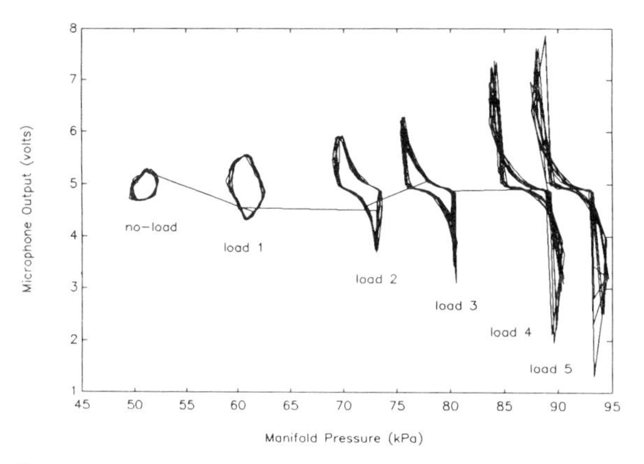

Figure 7 (Top) Microphone output voltage induced by cavity pressure fluctuations (bottom) for six load cases.

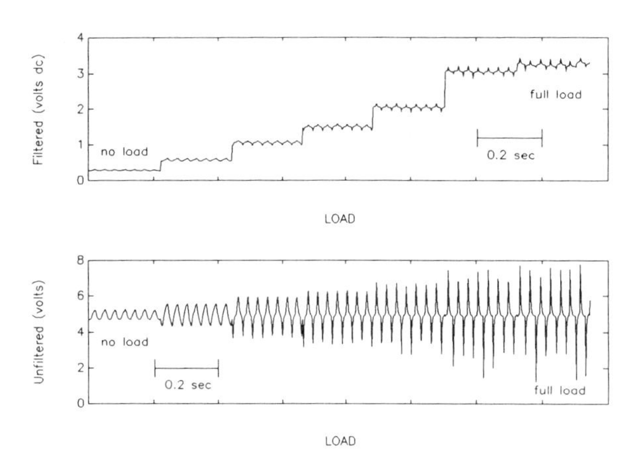

Figure 9 Rectified and filtered microphone output (top) shown with raw microphone signal (bottom) for six load cases.

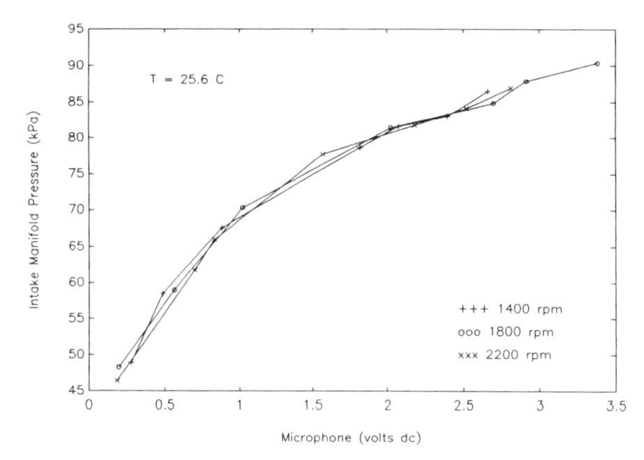

Figure 10 Calibration curves for the microphone dc output at three different speeds (seven load points each).

Figure 8 Phase plot of microphone voltage vs. manifold pressure for each load case. (These plots are representative of the voltage vs. strain characteristic of the PVDF film.)

determining the manifold pressure. The raw microphone signal shown in Figure 7 can be rectified and filtered to produce a dc voltage proportional to the manifold gauge pressure. This filtered signal is related to the manifold gauge pressure in Figure 9 for a range of steady-state operating conditions.

It is apparent from Figures 6,7 and 9 that the dc microphone output is not linear in pressure. Figure 10 shows a calibration curve for the filtered microphone signal to gauge pressure. The plot shows that there is a slight dependence on speed in this calibration.

Using a calibration similar to the one in Figure 10, Figure 11 (top) contains both the calibrated microphone output and the filtered absolute pressure transducer signal for a series of engine transients. Figure 11 (bottom) contains a plot of the percent difference between these two signals.

ERROR ANALYSIS

Comparison of the pressure values obtained from the microphone with those from the pressure transducer showed that the r.m.s. (root mean square) error was 1.6% of the reading (1.2% of full scale) with a standard deviation of 2.5%. Figure 11 (bottom) shows clearly the kind of errors that can be expected. Transient errors are large for low pressure measurements. Under steady conditions, and at higher pressures the errors are much lower, typically less than 1.5%.

Given the results above, we conclude that the microphone can provide a gauge pressure signal adequate for use in a speed-density algorithm since the steady values are accurate. Under transient conditions, while the errors appear large, the information necessary for transient compensation is provided by the microphone.

SUMMARY

It has been demonstrated that a single microphone element can be used to determine engine speed, steady and transient manifold gauge pressures in small engine applications. There are some nonlinearities in this sensor but these do not preclude its use. The microphone is a viable sensor which offers

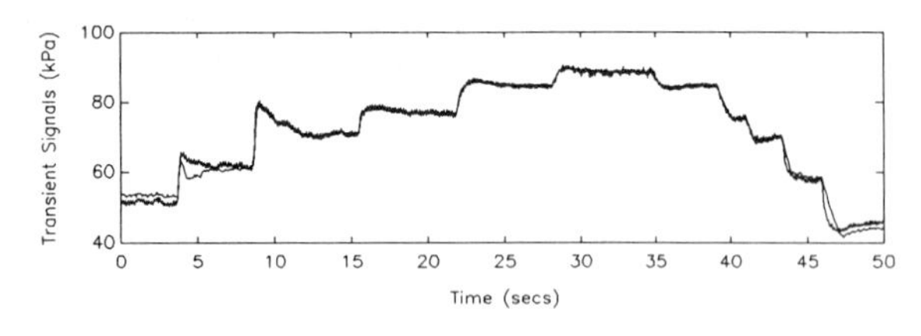

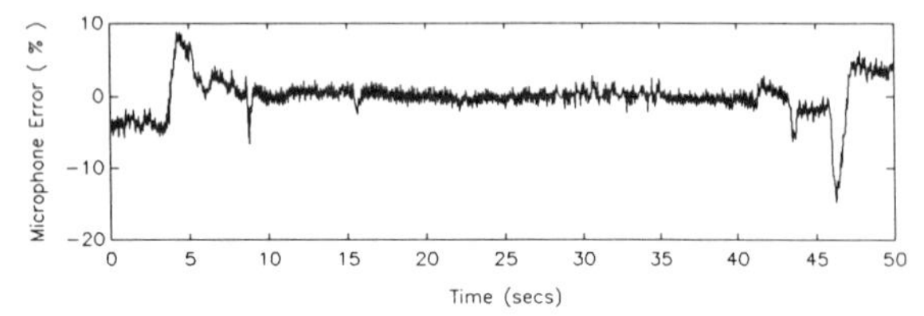

Figure 11 Comparison of absolute manifold pressure and microphone output voltage under transient conditions (top). Plot of percent error (bottom).

economic advantage over typical pressure transducers in this cost sensitive application.

ACKNOWLEDGEMENTS

Funding for this work was provided by Zenith Fuel Systems, Inc. of Bristol, Virginia and by the Virginia Center for Innovative Technology.

REFERENCES

Aquino, C.F.,"Transient A/F Control Characteristics of the 5 Liter Central Fuel Injection Engine", SAE 810494, SAE Trans., vol.90, 1981

Cook, R.D. and Young W.C., Advanced Mechanics of Materials, pp 179-185, Macmillan, New York, 1985

Eargle, J., The Microphone Handbook, pp 7-10, Elar Publishing Co., Plainview NY, 1981

Felger, G. and Plapp, G., "A New Single Point Fuel Injection System with Adaptive Memory Control to Meet the Most Stringent Emission Standards", I Mech E C221/85, 1985

Merhaut, Josef, Theory of Electroacoustics, pp 71-2, McGraw-Hill Int., New York, 1981

Wang,T.T.,Herbert,J.M., and Glass,A.M., Applications of Ferroelectric Polymers, Chapman and Hall, 1988

ICE-Vol. 15, Fuels, Controls, and Aftertreatment
For Low Emissions Engines
ASME 1991

CLOSED LOOP ENGINE CONTROL USING MULTIPLE MICROPROCESSORS

Rameshwar P. Sharma[1]
Department of Mechanical Engineering
Western Michigan University
Kalamazoo, Michigan

Subramaniam Ganesan[2]
Department of Computer
Science Engineering
Oakland University
Rochester, Michigan

Bharat Thacker
Universal Computer Applications
Southfield, Michigan

ABSTRACT

A generalized closed loop engine control system using multiple microprocessors has been designed and developed. Description of the hardware and software of the system and the test setup are given. Various parameters of the engine performance, such as, cylinder pressure, engine speed, engine load, fuel delivery, engine noise, airflow, and temperature are measured. A good correlation has been found in the control of the parameters in the laboratory. A summary of the on-going research in engine control is given.

INTRODUCTION

Real time control systems are very much influenced by the developments in advanced and high performance microprocessors and microcontroller integrated circuits. The microprocessors provide programmable implementation and present several advantages over the hardwired implementations. A few of the major advantages are: a) increased flexibility, b) increased reliability of the final system. c) reduced system implementation time, d) reduced system implementation costs,

A technique developed for stationary industrial gaseous fueled engines operating at Lambda values greater than one using an oxygen sensor is being developed and used for controlling very lean air fuel ratios and possibly to achieve the stringent low emission levels without a catalytic converter in most cases[1].

In this closed loop engine control system the other electronic performance controls can be easily incorporated for various engines because of the use of Reduced Instruction Set Computer (RISC) microprocessors[2].

More and more electronic circuits are used in the automobile every year to monitor various sensor signals and control the vehicle dynamically [3-6]. Microprocessors are used in engine control, idle speed control, transmission and gear box control [5-8]. Microprocessor based engine control systems constantly adjust fuel, ignition and emission controls. The result is better fuel economy, improved performance, and better drivability.

In certain luxury automobiles as many as thirty processors are used for various control applications. Some of the microcomputer systems are : multiprocessor based closed loop engine control system, microcontroller based anti- lock / traction control system, microcontroller based dynamic suspension control system, front panel display / status monitor system, Heat/ air temperature control system, and audio control using digital signal processors. The information communicated between these processors are: information from various sensors, computed results, control information, and status information. The communication rate should be fast enough for real time control operation. Special protocol or the controller area network (CAN) protocol can be used for the serial bus communication.

DEVELOPMENT AND TEST ENVIRONMENT

A block diagram showing serial communication between various microcomputer systems inside an example vehicle is given in figure 1.

(1) on leave at Research and Professional Services, Dearborn, Michigan

(2) on leave at Texas Instruments (TI), Houston Texas

A modern vehicle has several microprocessor based systems. In order to facilitate development and debugging and testing of these subsystems the designers have " Development, Debugging, and Test Tools (D^2T^2) ", which initially provide an interactive environment for the subsystem under consideration-" Engine Control".

D^2T^2 is capable of interfacing with several subsystems and can store various sets of results . User can specify or edit these results as parameters to the subsystem under consideration. A user can specify drag, noise etc.

In long term D^2T^2 can run several of these systems simultaneously and provide an on-line feedback from one subsystem to another.

Figure 2 shows a block diagram for a Digital Signal Processing (DSP) processor based closed loop engine control previously developed by the authors [7]. Closed loop engine control can tolerate external turbulences and provide optimum engine performance and fuel efficiency. In addition, the processor performs on-line real- time engine diagnostics.

The engine control unit sends control output signals based on the various sensor inputs. The common sensor inputs are:

- Air/Fuel ratio
- Engine speed (from distributor or crankshaft sensor)
- Engine load (from throttle position or vacuum lever sensors).
- Intake air temperature
- Intake air volume
- Barometric / manifold absolute pressure
- Vehicle speed
- Fuel Temperature
- In-cylinder pressure
- Transmission gear used
- Knock sensor (detects detonation)
- Oxygen sensor (detects the amount of unburned oxygen in the exhaust gas).

As compared to stationary engines, engines in vehicles have additional design considerations. Interfacing to other subsystems such as transmission is one example. Another consideration is emission control for modern vehicles. To meet the federal emission regulations, emission conversion is optimized by running the engine some what richer as shown in Figure 3A and 3B. From the same figure it is also clear that brake specific fuel consumption(BSFC) is compromised to achieve the required emission standards.

CONTROL ALGORITHMS

The electronic controller reads a set of sensors and generates necessary output signals to obtain the desired engine performance. Linearized description of an engine can be represented in a state space form as follows:

$$\dot{X} = AX + BU \quad (1)$$

$$y = CX \quad (2)$$

where

$\dot{X}$ = vehicle states rate vector, (nx1)
X = vehicle states vector, (nx1)
A = system matrix, (nxn)
U = actuators force vector, (mx1)
B = control input matrix, (nxm)
y = system output vector, (qx1)
C = output matrix, (qxn)

The control law is derived using the modern control theory. The control vector U is computed from $U=-K_cX +K_rR$, where K_c and K_r are the gain matrices and R is the reference input. An implementation of explicit adaptive servo control is further discussed in [8].

FUEL CONTROL

Figure 4 shows a schematic of the closed loop air fuel computer control. The computer controls fuel and other parameters (EGR/AIR) by responding to continuous exhaust gas oxygen sensing as an indicator of the air to fuel ratio in the engine. As a specific injector example, the each cylinder has a solenoid operated injector which sprays fuel toward the back of each intake valve. The length of time each injector is open governs the amount of fuel delivered. The electromagnetic valves are used to close and open the fuel pump and indirectly control the fuel injection system's performance. From a typical graph of solenoid pulse duration vs fuel delivery for various engine speed, we can obtain the expression for any given speed as:

At speed N_n

$$FD = m_n T + C_n$$

Where FD = fuel delivery
T = valve open time

From the input parameters which provide the air mass flow and the air to fuel ratio , the required fuel flow is computed. For this value at speed N_n, the slope m_n and constant C_n which are stored in the read only memory are used to compute the value T, the solenoid open time. For different speeds, N_n, the values of m_n and C_n are also different.

SPARK TIMING CONTROL

Figure 5 shows the closed loop control for spark timing. Measurement and analysis of engine pressure are done to obtain the crank angle vs cylinder pressure graph, as discussed in [9]. The knock sensor signal is used to retard spark timing or bleed off distributor vacuum.

The cylinder pressure transducers are used for measuring the in- cylinder pressure [10].

From the in-cylinder pressure the following features are extracted: location of peak pressure and knock detection. Inferred combustion parameters are: net heat release, combustion duration, combustion variability, Air/Fuel estimate, and MBT spark timing [11,12]. Sensor drift compensation are also achieved.

NOISE CONTROL

Noise control techniques are used mainly for the following four applications in a typical automobile.

1. The noise from the engine system is measured at a few selected frequencies. The amplitude of the noise signal is used as another parameter in the fuel / ignition closed loop control system. Use of these additional parameters will result in better fuel economy and also reduce the noise. A number of experiments are being done in the lab to find the effect of noise frequency / amplitude on the fuel economy. Development work on active noise cancellation techniques are being done.

2. It is desirable to determine the noise transmission paths from the front axle of a car to the driver or passenger under varying road / running conditions. The knowledge of the noise frequency and amplitude is useful in the selection and use of the components in the noise transmission paths. Figure 6 shows the correlogram of noise signals in an automobile. The microprocessor based system will compute auto and cross correlation functions and bandpass filter operations over the audio range and determine the frequency versus amplitude characteristics. The correlation and power spectral density functions are computed using the following expressions:

If a(t) and b(t) are sampled signals at time t. Let m be a variable to represent the delayed sampling interval. The value of m is, for this experiment ranges from 0 to $(2^7 - 1)$. When m=0, there is no delay between the samples of a(t) and b(t). When m=1, there is a 1 t time delay between the a(t) and b(t). Then the cross correlation function is given by:

$$R_{ab}(m\Delta\tau) = \frac{1}{N}\sum_{k=m}^{N+m-1} a(k\Delta\tau)\,b(k\Delta\tau-m) \qquad (4)$$

Where $\Delta\tau$ = sampling interval. N = the total number of samples taken for each signal. As shown in Figure 6, one signal may be measured by placing a transducer at the end of front axle. The second signal may be measured by placing a microphone inside the car. The peaks in the correlogram is due to various noise transmission paths. Each path introduces a different transmission delay. By eliminating the probable noise paths by suitable techniques the occurrence of the peak in the correlogram can be eliminated.

The power spectral density function, $P_m(r)$, gives the collective frequency and amplitude information taking into account the Hamming Window.

The power spectral density function is:

$$P_m(r) = 0.25P_0(r-1) + P_0(r) + 0.25P_0(r+1) \qquad (5)$$

$$\text{Where} \quad P_0(r) = \frac{\phi(0)}{2} + \sum_{i=1}^{h}\phi(i)\cos\left[\frac{2\pi ri}{2h}\right] \qquad (6)$$

$h = 128$ and $r = -1,0,1,2,3,...126$

and ϕ = auto correlation function.

To measure the frequency versus amplitude characteristics, the sampled signals are passed through overlapping digital bandpass filters. Digital bandpass filtering is obtained by a sequence of computations [13]. Overlapping bandpass filters are obtained by changing the center frequency in the above computations.

3. Measure the noise frequency and amplitude inside the car due to wind and other factors. Generate and transmit signals with the same frequency but with 180 degrees phase shift to nullify the noise. This is done to provide more comfort to the passengers. The noise is measured by placing microphones at strategic places inside the car. A equal number of speakers are used to transmit the phase shifted signal.

4. Measure the noise frequency and amplitude in the exhaust pipe. Generate and transmit signals with the same frequency but with 180 degrees phase shift to nullify this noise. This will reduce the exhaust system noise on the road and to the people outside the car.

These techniques are currently under development by the authors.

SYSTEM ARCHITECTURE

Multiple-microprocessor systems using the advanced microprocessors can provide an appropriate solution to the demand for additional computing power to support complex applications. A primary design objective of a multiple microprocessor system is enhancement of the system performance, throughput and real-time control. This, however, assumes that the computational task lends itself to partitioning into smaller tasks where one processor can be allocated to the execution of each task.

Dual Processor Architecture

Figure 7 shows the block diagram of one TMS320C25 DSP and a 68000 processor communicating through a dual-port common memory. This arrangement allows two separate processors running on independent clocks (asynchronously) to operate on the same data space. This architecture has been tested in the laboratory and is being used in the tests. The details about dual-port memory have been discussed by the authors in [7].

TESTS SUMMARY AND CONCLUSION

In the university laboratory testing and development work setup , all the above discussed engine parameters have been successfully tested in a single / multi cylinder engine [18].

For a vehicle control and better vehicle performance, the sensor signals from the engine speed, transmission, and air fuel are to be synchronized so that the vehicle performance is at the most feasible level for the possible fuel economy. Measurement of engine/ vehicle noise and other vehicle/ engine parameters have been done in the laboratory.

FUTURE DEVELOPMENTS

The future development of enhanced D^2T^2 will provide simultaneous running of several subsystems under test. This will provide online feedback among various subsystems. This essentially will simulate interaction of various subsystems as it would happen in a vehicle.

REFERENCES

1. D.W.Moss, and D.Wang, " Design and development of the Waukesha custom lean burn control system" ASME - ICE Fall Tech. Conf. Proceedings, Oct 15-18, 1989.

2. G.McClendon and C.Nampon, " Retrofit application of electronic performance controls and digital ignition to large bore, slow and medium speed engines", ASME - ICE Fall Tech. Conf. Proceedings, Oct 15-18, 1989.

3. K.N. Majeed, " Dual processor controller for vehicle suspension applications ", IEEE Trans. Vehicular Technology, Vol 39, No 3, August 1990, pp 271-276.

4. Engine and Drive line control systems published by SAE SP-739, March 1988.

5. H. Haanselmann, "Implementation of digital controllers- a survey" Automatica, Vol 23, 1987.

6. K.N. Majeed "Dual Processor automotive Controller" 1988 IEEE workshop on automotive applications of electronics IEEE Cat.No. 88TH0231, 19 Oct. 88, pp 39-44.

7. S.Ganesan, R.P.Sharma, " Engine control using DSP microprocessors", ASME - ICE Fall Tech. Conf. Proceedings, Oct 15-18, 1989.

8. M. Sunwoo, K.C.Cheok, S.Ganesan," An imple- mentation of explicit adaptive servo control with DSP based control system" Proceedings of the 19th annual Modeling and Simulation conference, Pittsburgh, May 1988.

9. D.R. Lancaster, R.B. Krieger, J.H.Lienesch, "Measurement and analysis of engine pressure data" SAE #750026.

10. C. Anastasia and G. Pestana, "A cylinder pressure sensor for closed loop engine control" SAE # 870288 SAE conference record, 1987.

11. R.P. Sharma and S. Ganesan," Microprocessor based controller for fluid System Application" Proceedings of Mini and Micro computer & application, Montreal 1985 page 158-161.

12. R.P. Sharma, "Design and development of electromagnetic fuel injection system," Proceedings of Western Michigan Conference on Mechanical Engineering, Kalamazoo, Oct 1984 page 370-380.

13. M.O.Ahmad, S.Ganesan, M.N.S.Swamy, " A real time programmable wave digital filter bank", Proceedings of IEEE int. Conf.

14. D. Hrovat etal, "Power Train Computer Control System" Automatic Control- World Congress 1987 (Oxford, UK: Pergamon 1988) Page 213-219.
15. P.W. Masding etal, "A Microprocessor controlled gearbox for use in Electric & Hybrid Electric Vehicles", Trans. Instr. Meas. Control (UK) Volume 10, No.4, page 177-186, July-Sept 1988.

16. S.R. Vishnubotla and S.M. Mahmud, A centralized Multiprocessor based control to optimize performance in complex vehicles" 1988 IEEE workshop on automotive applications of electronics (IEEE Cat.No. 88 TH 0231) 19 Oct 1988, Page 52-56.

17. T.J. Flis, "The use of microprocessors for electronic engine control" IEEE workshop on automotive applications of microprocessors, 1982, page 3-19.

18. Sharma, R.P., "Parametric Study of the FORD PROCO Fuel Injection System and Fuel Delivery Line", 1983 ASME Publication No. 83-WA/DSC-27.

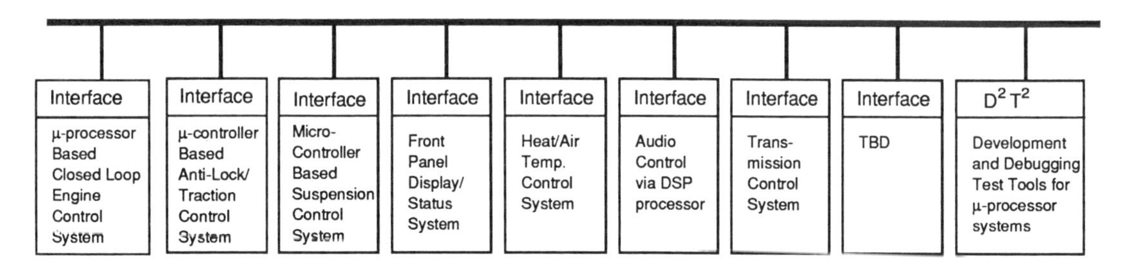

Figure 1 **Block Diagram Showing** Serial Communication between various Microcomputer systems inside example vehicle. (University Lab Set-up)

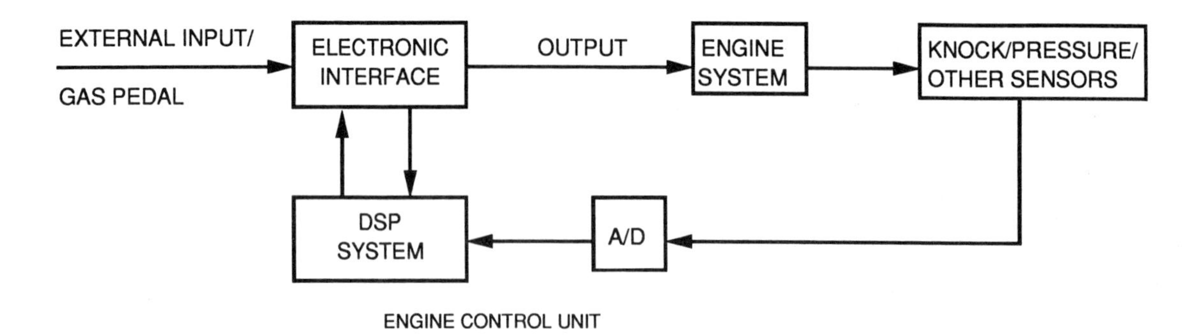

INPUT: CRANK POSITION, ENGINE SPEED, OXYGEN SENSOR, A/C SWITCH, MANIFOLD ABSOLUTE PRESSURE, AIR TEMPERATURE, AIR MASS FLOW, BATTERY POWER, COOLANT TEMP, KNOCK SENSOR

OUTPUT: FUEL INJECTOR CONTROL, SPARK TIMING, IDLE SPEED CONTROL, EGR VALVE POSITION CONTROL

FIGURE 2: CLOSED LOOP ENGINE CONTROL

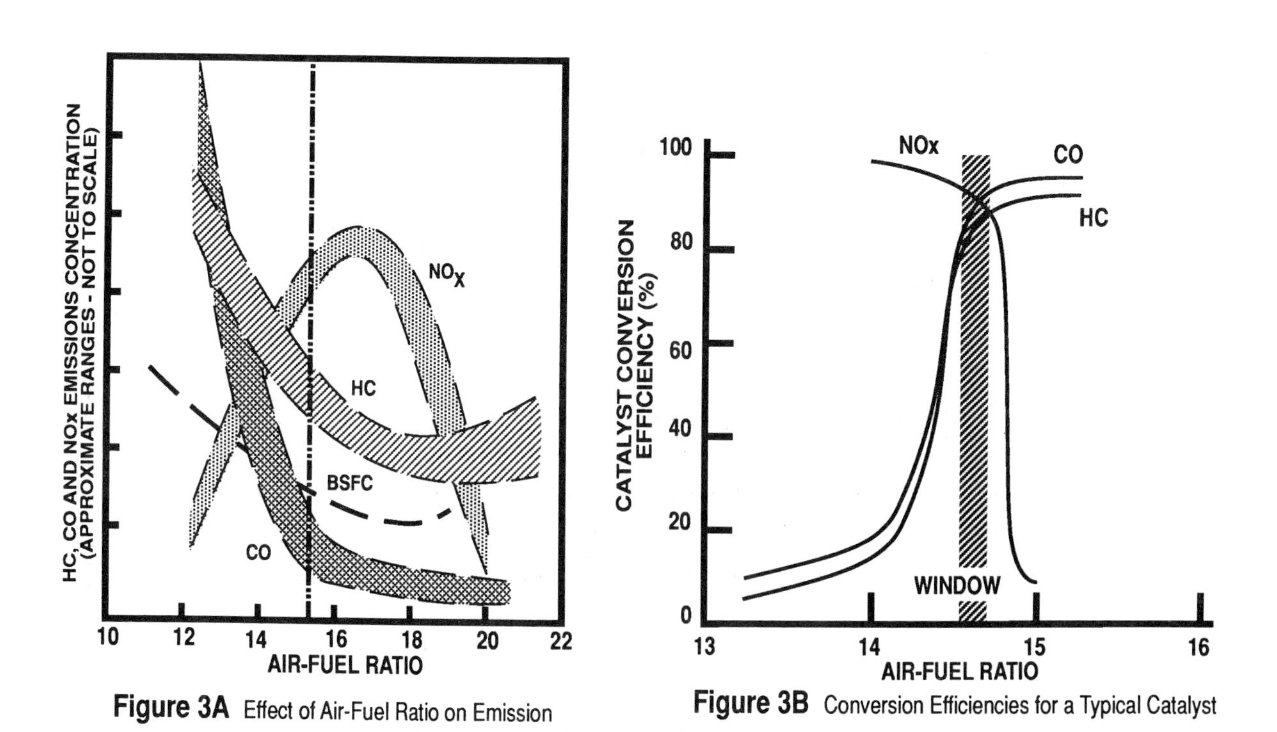

Figure 3A Effect of Air-Fuel Ratio on Emission

Figure 3B Conversion Efficiencies for a Typical Catalyst

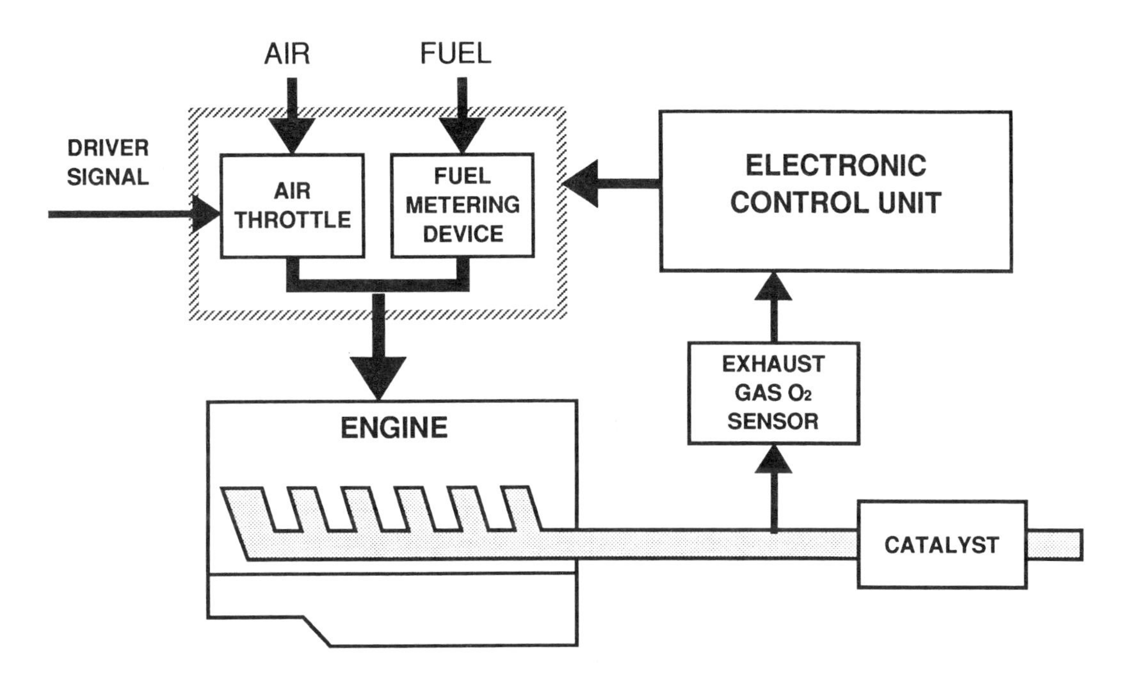

Figure 4 Schematic Diagram of a Closed-loop Control System

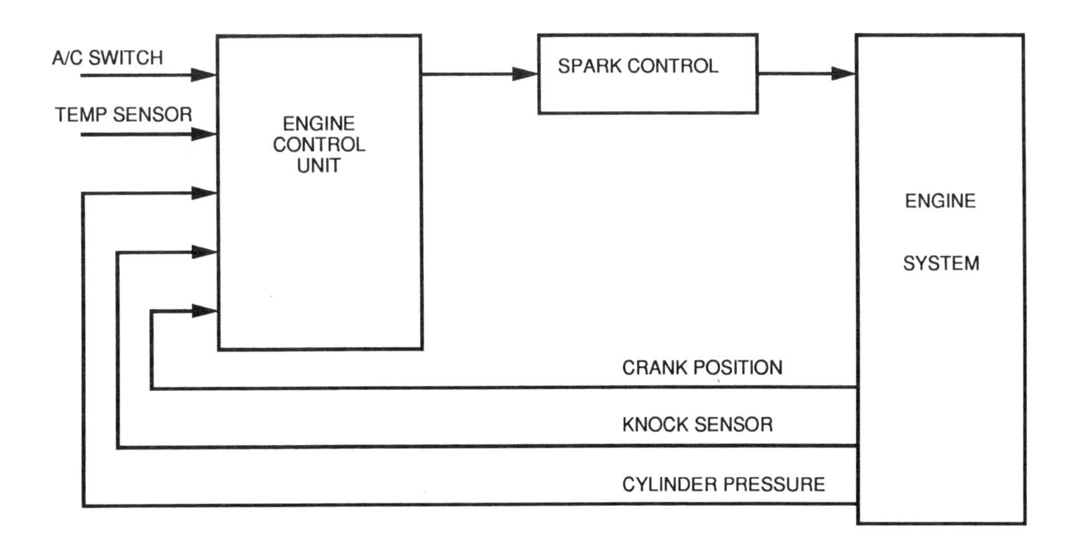

FIGURE 5: CLOSED LOOP CONTROL FOR SPARK TIMING

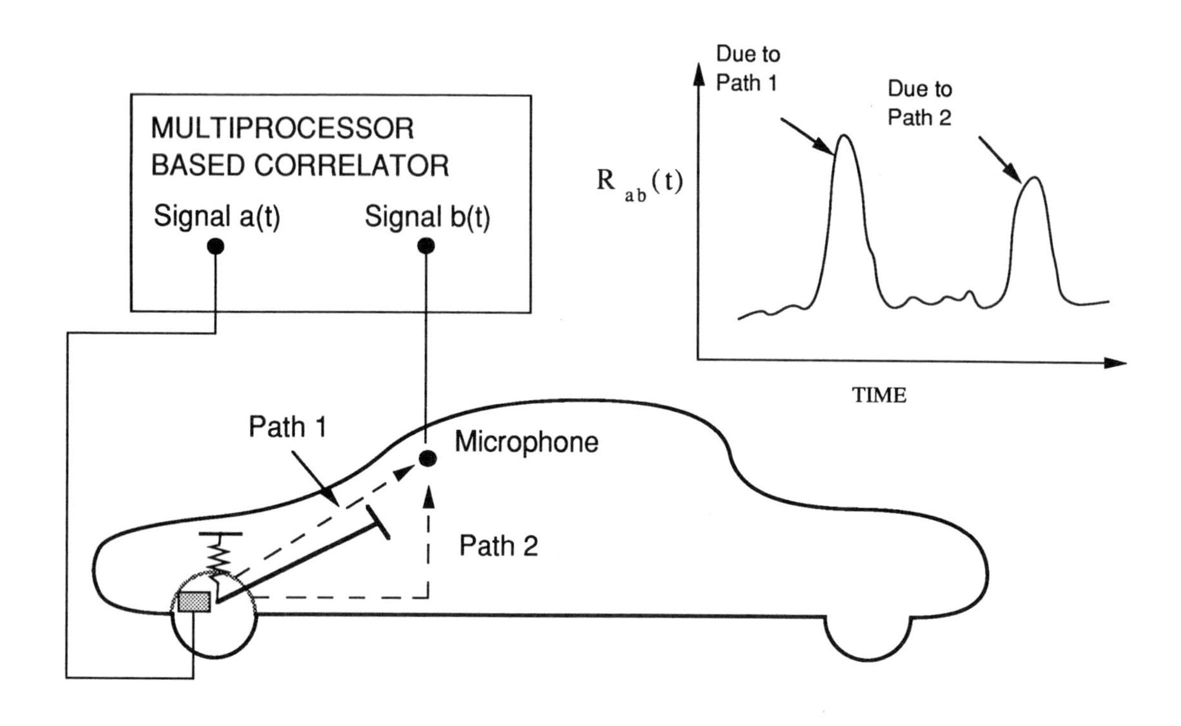

FIGURE 6: CORRELOGRAM OF NOISE SIGNALS IN AN AUTOMOBILE

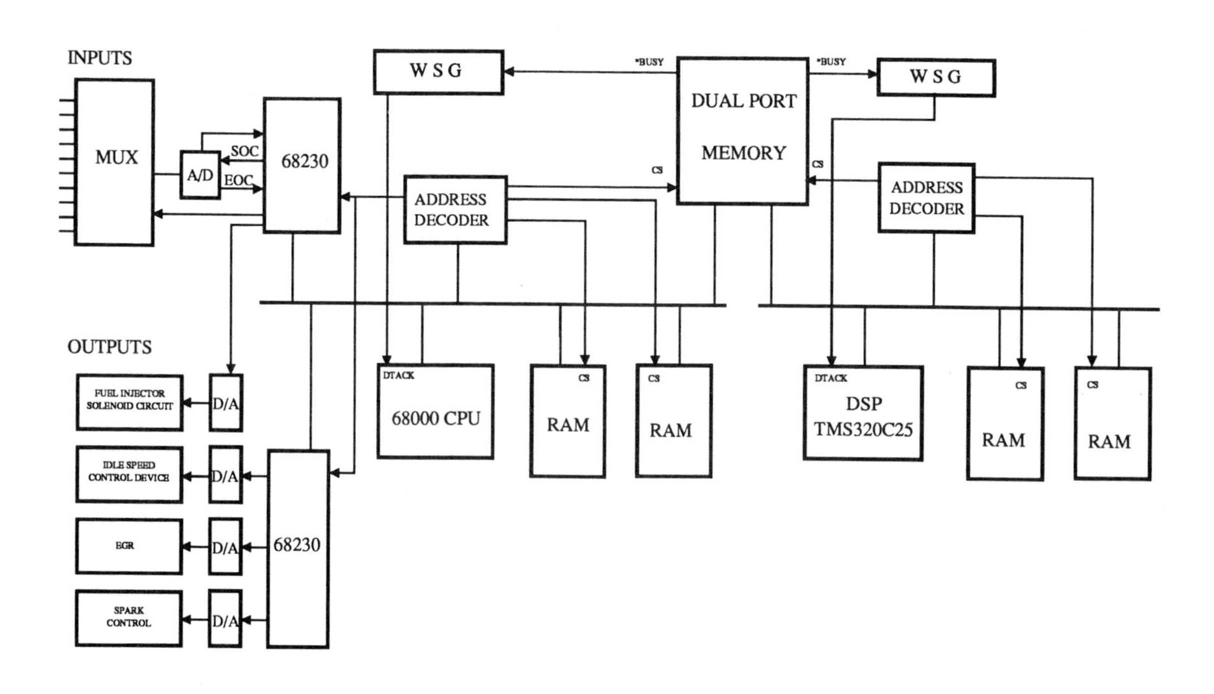

FIGURE 7: ENGINE CONTROL SYSTEM ARCHITECTURE

ICE-Vol. 15, Fuels, Controls, and Aftertreatment
For Low Emissions Engines
ASME 1991

NATURAL GAS FUELING OF A CATERPILLAR 3406 DIESEL ENGINE

G. E. Doughty, S. R. Bell, and K. C. Midkiff
Department of Mechanical Engineering
University of Alabama
Tuscaloosa, Alabama

Abstract

A Caterpillar 3406 turbocharged diesel engine was converted to operate in a natural gas with diesel pilot ignition mode and was evaluated for performance and emission characteristics for both diesel and natural gas operation. Full load power was achieved with natural gas fueling without knock. Similar fuel efficiencies were obtained with natural gas fueling at high loads, but efficiencies were lower for low loads. Bosch smoke numbers were reduced by over 50 percent with natural gas fueling for all cases investigated. NO_x emissions were found to be lower at low loads and at high speeds under high load. CO emissions were significantly increased for natural gas fueling while CO_2 concentrations in the exhaust were reduced for natural gas fueling.

Introduction

Petroleum, since the turn of the century, has been an inexpensive and readily available energy source. Experts for some years now have predicted the end of this era. There may not be agreement as to exactly when the petroleum supply will be depleted, but the concern is great enough to warrant research in alternative fuels. A successful alternate fuel should be inexpensive, abundant, and the products of combustion of the fuel must be environmentally acceptable or be such that they can be cleansed to acceptable levels.

Concern with exhaust emissions pollution from internal combustion engines has become a major driving force in new engine designs and new fuel development for engines. Environmental problems have been particularly severe in urban areas where high concentrations of vehicles and people result in the exposure of a large population to high levels of pollutant emissions. Current enacted and proposed legislation by several governmental bodies across the United States is aimed at addressing this environmental concern. For pre-1990 diesel engines, the EPA allowed 10.7 g/bhp-hr of NO_x emissions. This limit was reduced to 6.0 g/bhp-hr in 1990 and to 5.0 g/bhp-hr in 1991. In January of 1988 and for the first time, the EPA began regulating particulate emissions from diesel engines. The initial standard for particulates is 0.6 g/bhp-hr with a final proposed standard of 0.1 g/bhp-hr beginning in 1994. [1,2] These emission standards are extremely demanding and potentially extend beyond the current marketplace technology for diesel engines. New and innovative concepts with relatively short developmental times must, therefore, be sought if their impact is to be realized in the market in a timely manner.

The use of alternative fuels and in particular natural gas has been identified as a potential design choice for engines. [1,2] The performance of natural gas engines has been investigated with promising results at The University of Alabama and by others. [3-8] The emissions aspects are less well investigated, but preliminary results from testing are encouraging for the particulate and NO_x emission levels.

An advanced and promising concept of natural gas fueled engines is the lean burn engine. The lean burn engine, as the name implies, operates with a significant fraction of excess air. The primary potential advantages offered by the concept are lower local temperatures in the cylinder which yields lower NO_x emissions, less potential for knock and the potential for improved thermal efficiencies. Conventional spark ignited gas engines operate close to the stoichiometric fuel/air ratio which results in higher peak local temperatures which increases the chances of dissociation reactions for forming NO_x and often results in incomplete combustion. While conventional fueled diesel engines operate at an overall lean condition, the in-cylinder fuel injection results in actual burning occurring at conditions ranging from rich to lean, thereby producing products characteristic of rich burning (particulates) and of stoichiometric burning (NO_x). The application of natural gas fueling to a diesel type engine is a lean burn concept if the fuel is either aspirated or injected prior to entering the cylinder so that there is a near premixed mixture at ignition. Ignition of lean methane mixtures is difficult to achieve and can result in incomplete combustion or total misfire [9]. For ignition to be successful the energy release rate in the early stages of ignition must be greater than losses from the ignition flame kernel. If not, the flame extinguishes prematurely. For lean mixtures, the energy release per unit volume is less because the fuel charge is diluted with excess air. This is a major concern for lean burn engines and is a concern which warrants careful study. Concepts suggested for achieving stable ignition include using a pilot diesel charge, high energy spark, plasma jets and stratified charge/spark designs. [9-11] The objective of this work was to investigate the emissions and performance characteristics of a commercial diesel engine being operated on natural gas with pilot diesel for ignition. For the results reported, the engine design and operating conditions were not specifically modified to minimize emissions for operations on natural gas. However, the engine did exhibit strong emissions dependency on those operational parameters varied. The following sections of this report include a discussion of the testing facilities and discussion of results and conclusions.

Facility Description

The engine used in this study was a Caterpillar 3406 diesel. The engine was factory equipped with a turbocharger and retrofitted with an intercooler to allow operation using natural gas without encountering knock. The basic engine characteristics of the test engine are summarized in Table 1. [6]

A drawing of the engine test facility is shown in Figure 1 and includes the major equipment and support systems for the facility. A coolant tank was constructed of stainless steel to hold coolant water for the test engine. The tank was equipped with a water inlet and an overflow drain such that the tank temperature could be controlled by the flowrate of the water. The tank was also equipped with a thermocouple and digital readout so that coolant temperature could be monitored. All engine tests were conducted with the coolant water at approximately 180F. The intercooler used city water to cool the intake air-fuel mixture so that natural gas might be used as a fuel without knock. Inlet water temperatures at the intercooler and the coolant tank were approximately 67F.

The natural gas used was pipeline natural gas from the local gas utility and was introduced into the intake air stream just

Table 1 Engine Specifications

Cylinders	6, inline
Bore	5.4 [in]
Stroke	6.5 [in]
Displacement	893 [in^3]
Compression Ratio	14.5
Combustion System	Direct Injection
Maximum Engine Speed at Full Load	2100 [rpm]
Rated Brake Power	280 [hp]
Fuel Injection	Variable Time

Table 2: Chemical Composition and Analysis of Natural Gas

Component	Formula	Volumetric Analysis (%)
Hexanes +	C_6 +	0.13
Propane	C_3H_8	0.57
Iso-Butane	i-C_4H_{10}	0.11
Normal Butane	n-C_4H_{10}	0.15
Iso-Pentane	i-C_5H_{12}	0.05
Normal Pentane	n-C_5H_{12}	0.06
Carbon Dioxide	CO_2	0.28
Ethane	C_2H_6	3.29
Oxygen	O_2	< 0.01
Nitrogen	N_2	0.96
Methane	CH_4	94.39
Carbon Monoxide	CO	< 0.01

Compressibility Factor Z = 0.99785
Specific Gravity (Air = 1.0) Ideal = 0.5910
HHV @ 60°F and 14.73 psia Dry Gross Ideal = 1047.0 BTU/ft^3
Specific Gravity (Air = 1.0) Saturated = 0.5921
HHV Saturated = 1031.0 BTU/ft^3

prior to the turbocharger. Natural gas flowrates were controlled using a manual, variable area, fine control needle valve. A gas analysis was preformed to determine the gas composition and theoretical heating value of the natural gas, see Table 2. The higher heating value of the natural gas was taken to be the sum of the weighted averages of the higher heating values of each constituent. To determine the lower heating value of the natural gas the heat of vaporization of the water contained in the products was subtracted from the theoretical higher heating value. This gave a lower heating value for the gas of 20600 BTU/lbm.

Engine intake air was filtered and passed through a laminar flow element and surge drum before entering the engine, see Figure 1. For tests where the air to fuel ratio was varied, the air flowrate was controlled using a butterfly valve upstream of the natural gas inlet and the turbocharger. Diesel fuel was stored in a 55 gallon drum that was higher in elevation than the fuel pump on the engine. The diesel fuel was supplied by the university and had a cetane number of approximately 49. The lower heating value of the fuel was estimated to be 18280 BTU/lbm. [12] The diesel fuel flowrate for pilot ignition was measured using a positive displacement flowmeter which provided an analog output for the data acquisition system. Several temperatures were monitored in the engine including each exhaust port, pre- and post-turbo, inlet air, coolant in and coolant out. Pressure measurements included laminar flow element differential, pre- and post-turbo and exhaust manifold.

The test dynamometer was a directly coupled, water cooled, traction brake. It was equipped with a digital read out of engine speed, torque, and power. A strain gauge amplifier was placed in parallel with the dynamometer strain gauge and a frequency to voltage circuit was placed in series with the magnetic pick-off on the dynamometer. These modifications allowed engine speed and torque to appear on the dynamometer digital readout and to be recorded by the computer data acquisition system simultaneously.

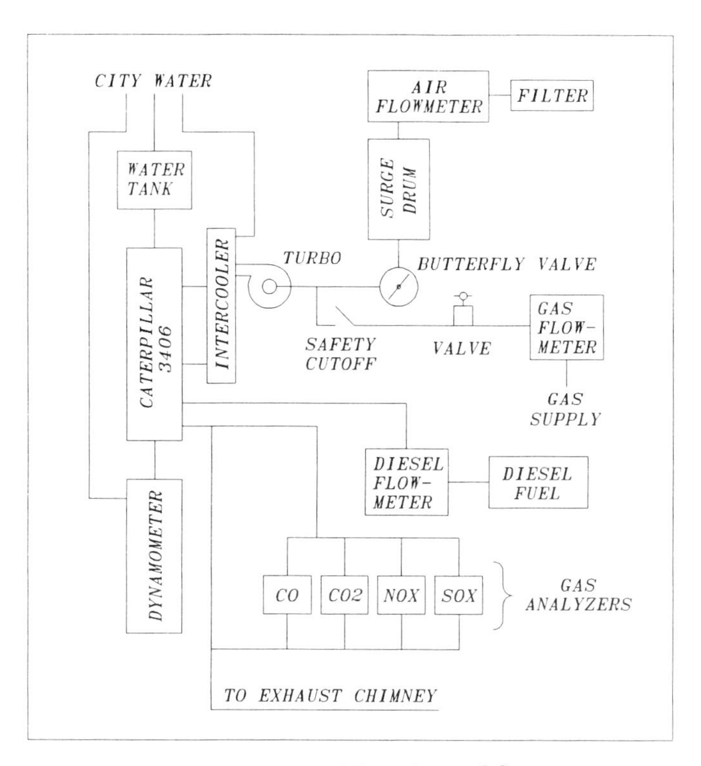

Figure 1 Schematic of Experimental Setup

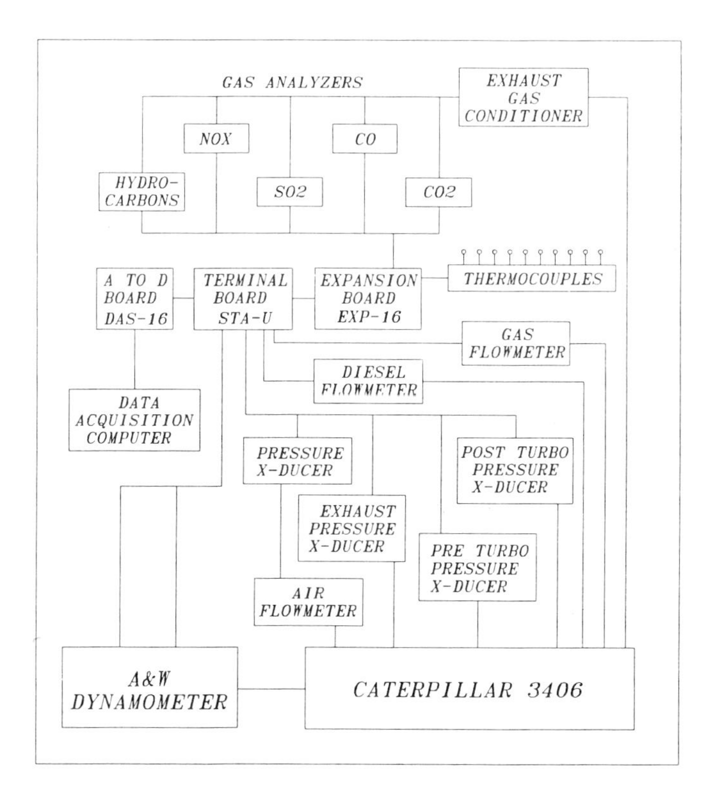

Figure 2 Schematic of Data Acquisition System

Continuous sampling exhaust gas analyzers were used to measure the levels of NO_x, CO, CO_2, and SO_2, in the exhaust. Sample lines were connected to the exhaust manifold directly behind the turbocharger and the gas was conditioned using a Thermo Electron Corporation model 600 sample gas conditioner. The conditioner passes exhaust gas through a filter and a 37 degree F bath to cleanse and dehumidify the gas. The gas is then passed through a molecular dryer for further dehumidification and then to the analyzers. Each instrument is equipped with a 0-100 mV output. The voltage output from each instrument is proportional to the concentration of pollutant in the exhaust. Particulates concentration in the exhaust were monitored using the Bosch smoke number method. Samples were taken in the exhaust pipe approximately 4 feet from the exhaust manifold. A Bosch sample pump was used to pass a fixed amount of exhaust gas over a filter to trap the particulates. A Bosch smoke meter was then used to assign a smoke number for each filter paper.

The data acquisition system used was a Swan XT10 computer equipped with a Metra Byte Corporation Das-16 high speed analog to digital interface board. The operating parameters including temperatures, pressures, flowrates, engine speed, torque, and pollution content (NO_x, CO, CO_2, SO_2) of the exhaust gas are monitored and recorded in data files using this system. The software program converts bit values returned by the DAS-16 to data readings such as torque and fuel flowrate and calculates from these power, fuel efficiency, and emissions. A schematic of the instrumentation and data acquisition is shown in Figure 2.

Results and Discussion

Two separate groups of tests were conducted in this study. In the first group of tests, the effects of varying the speed and load of the engine on the emissions and performance were investigated. The engine load was varied in percentages of full load in 25 percent increments as engine speed was varied in 200 rpm increments. Figure 3 illustrates the power curves for each load. This graph represents actual data from a diesel baseline test and compared closely to manufacturer's data. The power results for natural gas and diesel runs were similar as the natural gas flowrate was adjusted to produce the diesel performance at each speed and load. In all cases, the diesel pilot flowrate was held near 15 percent of full load flowrate on a mass basis. For this paper, only the full and one-half load data are presented graphically. Trends exhibited for the other loads closely followed the full and one-half load cases.

In the second group of tests the intake air was restricted using the valve located upstream of the turbocharger. This allowed investigation of the effect of air to fuel ratio on engine performance and exhaust emissions. Engine speed was held constant for each test while loads and air to fuel ratios were adjusted. The tests were conducted at three engine speeds, 1300 rpm, 1700 rpm, and 2100 rpm, although only the 2100 rpm data is presented in figures.

Figures constructed with all recorded data were difficult to read due to overlapping. Therefore, graphical representation of test data presented in this chapter is given in the form of best fit curves. Error bars representing the standard deviation are

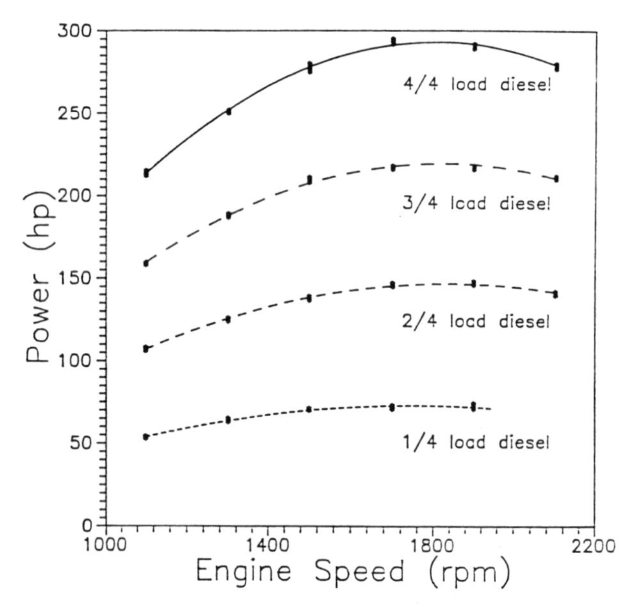

Figure 3 Power as a Function of Engine Speed for Full, Three-Quarters, One-Half, and One-Quarter Loads

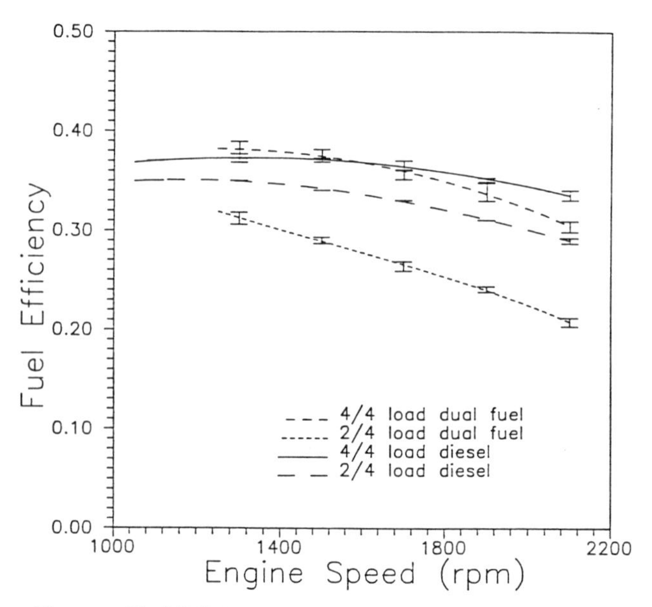

Figure 4 Fuel Efficiency as a Function of Engine Speed for Full and One-Half Loads

included in graphs where feasible in order to give an indication of the data deviation. For all cases, approximately 12 data points at each operating condition were used to determine the curves and the standard deviation.

Diesel curves represent the combination of 2 sets of data taken on separate days after recalibration of the equipment. Dual-fuel curves that give data as a function of load and rpm represent the combination of 3 sets of data taken on separate days after recalibration of the equipment. Dual-fuel curves that give data as a function of fuel/air equivalence ratio and load represent the combination of 2 sets of data taken on different days after recalibration of the equipment.

Performance Results

The performance results presented includes the brake fuel efficiency and fuel-air equivalence ratio. Other parameters measured include brake torque, brake power and volumetric efficiency. Fuel conversion efficiency was taken to be equal to the brake power divided by the total injected energy rate (pilot diesel + natural gas). Because fuel heating values are different for diesel and natural gas brake specific fuel consumption comparisons were not made. However, these can be calculated from the presented material.

Figure 4 shows the brake fuel conversion efficiency as a function of engine speed for full and one-half load for both diesel and natural gas fueling. The full load fuel efficiency closely matches the diesel full load being slightly higher at speeds below 1500 rpm and lower at speeds above 1500 rpm. As speed of the engine is increased, the combustion event is prolonged (in terms of crank angle duration) and moves further into the expansion stroke. The pilot injection timing is analogous to spark timing in that it determines the start of the combustion event. For natural gas fueling, no modifications to the injection timing of the pilot diesel were made over standard diesel timings. It is expected that the performance as a function of engine speed results could be optimized using pilot timing for natural gas. This, however, was not done in this study. For one-half load, the natural gas fuel conversion efficiencies were lower than the corresponding diesel efficiencies by 8 to 30 percent over the operating speeds investigated. Again, this was partly attributed to the non-optimized pilot timing. Cylinder pressure data taken previously for this engine for natural gas fueling showed the peak pressure occurring between 5 and 10 crankangles later (depending on load and speed) than the corresponding peak pressure for diesel fueling. In some cases a double hump on the pressure trace could be seen being explained as combustion of the pilot diesel charge followed by the methane combustion. [6] A Caterpillar 3208 engine was also previously tested using natural gas and similar dual peak pressure traces were found under some conditions. With a 7.5 crankangle advance in the 3208, a 10 percent increase in fuel conversion efficiency at higher speeds was seen for natural gas fueling. For diesel fueling, the same advance decreased the fuel efficiency by about 45 percent at higher speeds. [6] For the 3406 engine tested here, it is envisioned that advancing the timing for natural gas fueling would lead to performance improvement. As discussed later, emissions optimization would likely require timings different to best performance timings.

As engine load is further reduced, the fuel efficiency continues to drop. The worst case recorded was one-quarter load at 1900 rpm with the fuel efficiency of diesel being approximately 24 percent and natural gas being approximately 17 percent. As the load is decreased, the fuel-air equivalence ratio in the engine is leaned. The strong dependence of flame speed for methane-air mixtures (as well as other combustion characteristics, such as required ignition energy) on equivalence ratio impacts the engine performance. [13-15] As a methane-air mixture is leaned from stoichiometric the flame speed is decreased and for a fixed ignition timing will lead to longer or delayed combustion. The delayed peak cylinder gas pressure relative to TDC and lean quenching of the methane-air mixtures in the periphery of the combustion chamber are probable explanations for the reduction of efficiencies. Although not included in this study, it would be expected that an increase in unburned methane would be found at the light loads.

An important parameter for consideration is, therefore, the fuel-air equivalence ratio which is defined as the stoichiometric air to fuel mass ratio divided by the measured air to fuel mass ratio.

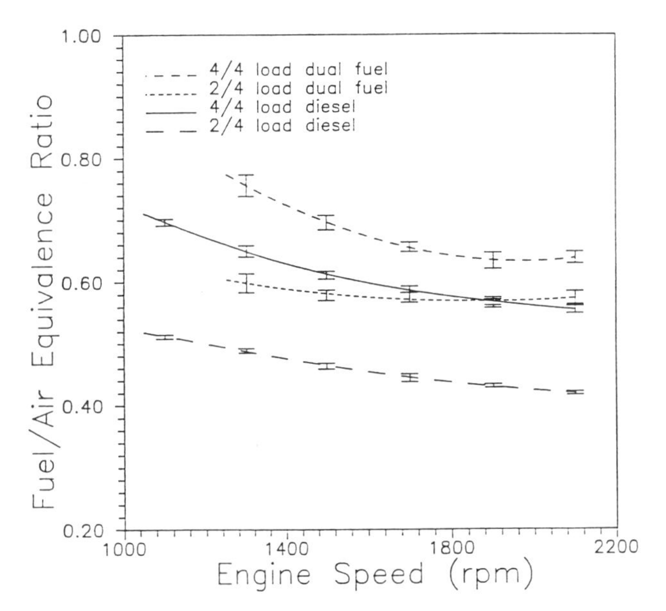

Figure 5 Fuel-Air Equivalence Ratio as a Function of Engine Speed for Full and One-Half Loads

Figure 5 shows the equivalence ratio as a function of engine speed for full and one-half load for both diesel and natural gas fueling. As shown, the natural gas fueling compared to diesel fueling leads to slightly higher (less lean) equivalence ratios for a given load and speed condition. This occurs for two reasons. First, the natural gas is aspirated into the engine where it mixes with air thereby displacing some portion of air which could have moved into the cylinder. As less air is inducted, the equivalence ratio increases. Secondly, as load is decreased the engine is less efficient using natural gas (as discussed earlier) and more natural gas must be added to produce the fixed load-speed condition. The increased fueling then increases the equivalence ratio of the engine.

For natural gas fueling, increasing speeds and decreasing loads resulted in leaner mixtures. As shown in Figure 4, this also led to lower fuel conversion efficiencies. Worth noting here is that for diesel fueling the fuel and air are not in a premixed form for combustion and the equivalence ratio is an overall ratio. Actual combustion occurs in fuel rich to stoichiometric to fuel lean regions. Therefore, direct comparisons between the diesel equivalence ratios and natural gas equivalence ratios can be misleading as one is natural gas-air and the other is diesel sprayed into air.

The fuel-air equivalence ratio impacts the combustion process which in turn affects engine performance and emission characteristics of the engine. To investigate the effect of equivalence ratio on fuel efficiency, the inlet air to the engine was throttled thereby allowing control of the equivalence ratio. Figure 6 shows the fuel conversion efficiency as a function of fuel-air equivalence ratio for natural gas fueling at full and one-half load at 2100 rpm. Also shown in the figure for comparison are the full and one-half load points for the base diesel cases. For full load natural gas fueling, only modest improvement of efficiency could be attained by increasing the equivalence ratio. For half load, efficiency improvement was more pronounced, increasing from 21 percent at .51 equivalence ratio to 25 percent at about .80 equivalence ratio. The best fuel efficiencies appear to occur at equivalence ratios near .75 to .80 for these cases. For both loads, the natural gas fuel efficiencies are lower than the corresponding diesel case. Again, it is expected that optimizing the pilot charge timing for best fuel efficiency would lead to better natural gas fuel efficiencies. As the equivalence ratio is decreased, the flame speed of the leaned mixture decreases which results in lowered engine performance. As the equivalence ratio is increased a point would eventually be reached where further increase leads to lowered fuel efficiency. For

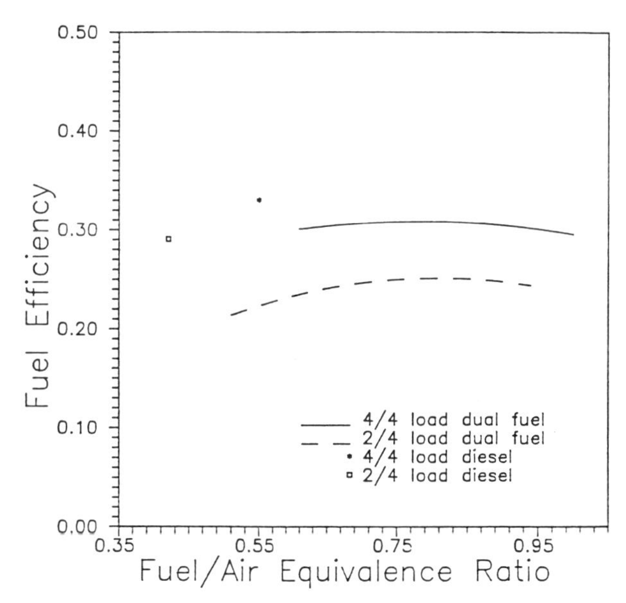

Figure 6 Fuel Efficiency as a Function of Fuel/Air Equivalence Ratio for Full and One-Half Loads and 2100 rpm

both load cases in Figure 6, a slight downturn in efficiencies was seen at equivalence ratios above about 0.9. Two possible explanations follow. First, it is expected that a best pilot injection timing for a given mixture equivalence ratio exists. In these results, the timing was not changed and therefore, the performance would decrease from a best value as the equivalence ratio is changed. Second, as the equivalence ratio is increased, to near or above 1, it is more difficult to achieve adequate air to all fuel locations in the cylinder leading to longer combustion times and possibly unburned fuel. This would also result in lowered efficiencies.

In summary, for performance, the engine was operated without knock with gas substitution percentages of about 85 percent on a mass basis. Full load power was easily achieved at all operating speeds. At lower loads a decrease in fuel efficiency was found, although adjusting the fuel-air equivalence ratio can moderate this effect. It was determined the pilot timing should be further investigated and hydrocarbon emissions data would be useful in determining partial burn limitations.

Emission Results

NO_x emission concentrations measured in the exhaust gas as a function of engine speed for full and one-half load for diesel only and natural gas fueling are shown in Figure 7. At full load and for speeds greater than about 1800 rpm natural gas fueling reduced the NO_x emission concentrations from the engine compared to the full load diesel fueling NO_x concentrations. Below 1800 rpm, NO_x concentrations are increased over diesel fueling levels for full load natural gas fueling. For half-load conditions the NO_x concentrations are lower for natural gas fueling for speeds above about 1400 rpm and slightly higher than the diesel levels for lower speeds. In both load cases, the general trend is decreasing NO_x concentrations with increasing speed. These results are consistent with the fuel-air equivalence ratio data presented earlier which showed natural gas-air mixture leaning with increasing engine speed. Mixture leaning for the near homogeneous natural gas cases would lead to lower incylinder temperatures and thus, decreased NO_x formation. Also, the leaner mixtures would result in slower flame propagation also resulting in lowered peak incylinder temperatures. The fuel efficiency data presented earlier is also consistent with this explanation with an overall decreasing trend with increasing speed. The non-optimized pilot timing (with respect to fuel efficiency) leads to lowered performance. However, the NO_x engine characteristics were improved. This is interesting in that it points to the trade-off between emissions and performance.

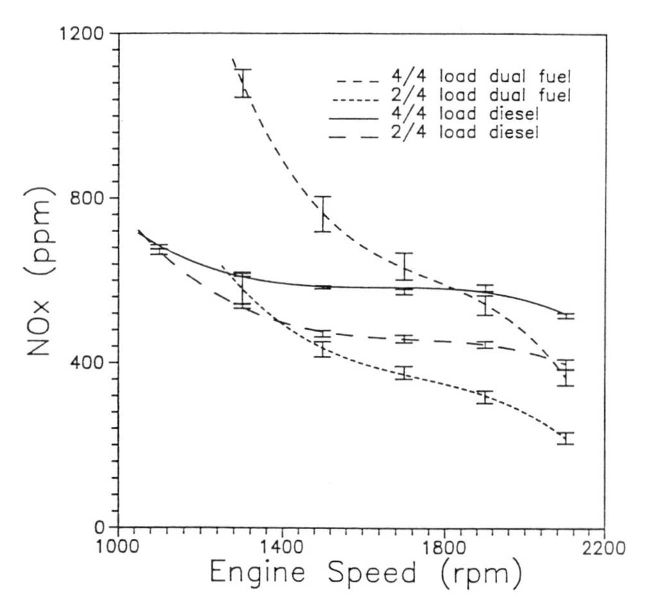

Figure 7 NO_x Emission as a Function of Engine Speed for Full and One-Half Loads

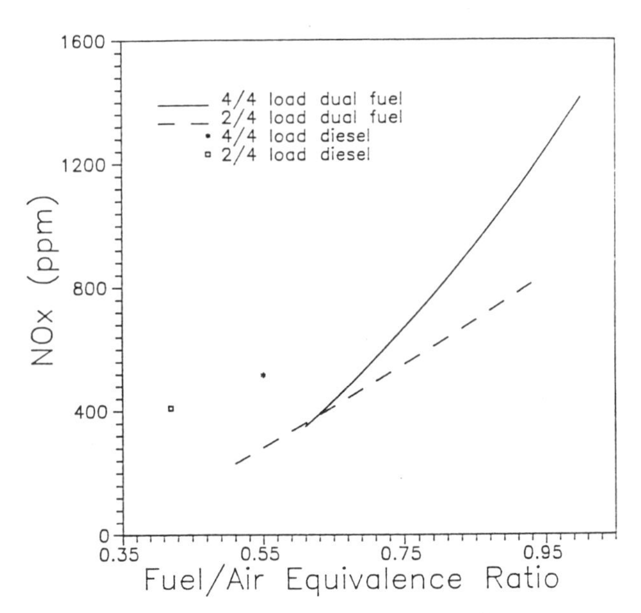

Figure 8 NO_x Emission as a Function of Fuel/Air Equivalence Ratio for Full and One-half Loads and 2100 rpm

Figure 8 shows the effect of varying the fuel-air equivalence ratio on NO_x concentrations for full and one-half load natural gas fueling at 2100 rpm. Again, the corresponding diesel only points are noted on the figure for reference. The left limit for the full and one-half load data lines corresponds to no throttling of the intake air and result in the leanest mixture conditions and lowest NO_x concentrations. As the air is throttled, less air is used to charge the cylinder and thus a more fuel rich mixture occurs. As would be expected higher equivalence ratios yield higher incylinder temperatures and thus, higher NO_x concentrations. The NO_x concentrations are also affected directly by the change in dilution air to the engine. That is, as excess air is decreased, less dilution occurs thereby increasing the NO_x concentrations. For half-load, increasing the equivalence ratio from .5 to 1.0 would reduce the number of product moles by approximately half. For a given NO_x production, this would increase the concentration of NO_x by 2 or double the NO_x. The measured NO_x concentration changes by a factor of about 4 for this equivalence ratio range and thus, about one-half of the NO_x change can be attributed to dilution changes. For full load conditions, dilution appears to play a less significant role in the NO_x concentration levels.

Figure 9 shows the concentration of CO in the exhaust as a function of engine speed for full and one-half load conditions for both diesel and natural gas fueling. For all speeds, natural gas fueling resulted in higher CO emissions. This is consistent with the fuel efficiency results and explanations for non-optimized pilot timing and flame quenching or partial burning. It is expected that the elevated CO concentrations would be accompanied by higher unburned hydrocarbons. Although not presented graphically, varying the fuel-air equivalence ratio allowed CO emissions to be lowered, however, the best CO emission levels for full and one-half loads (occurring at an equivalence ratio of .75 to .80)

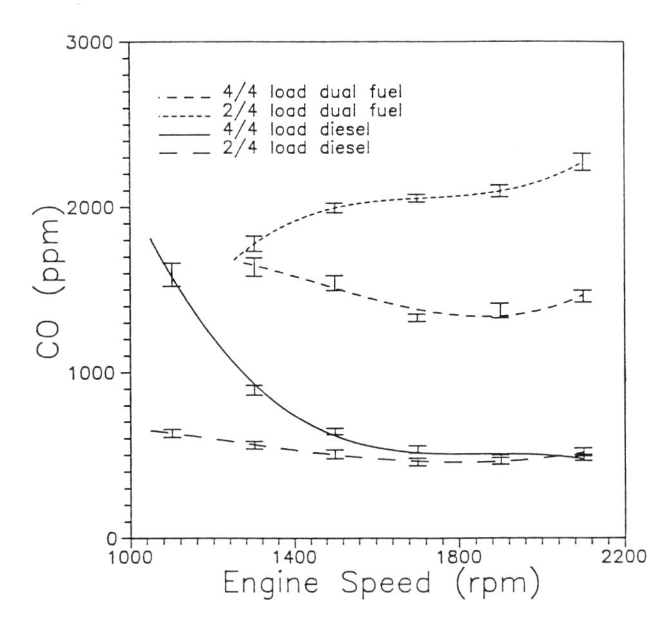

Figure 9 Carbon Monoxide Emission as Function of Engine Speed for Full and One-Half Loads

were still higher than the corresponding diesel only levels. It is expected that pilot charge timing could also be used to control CO concentration levels.

The CO_2 concentration in the exhaust gas was lowered for natural gas fueling. These results are shown for full and one-half loads in Figure 10. The decrease in CO_2 concentrations results from the higher H to C ratio of natural gas which directly impacts the number of moles of each product constituent. As the H to C

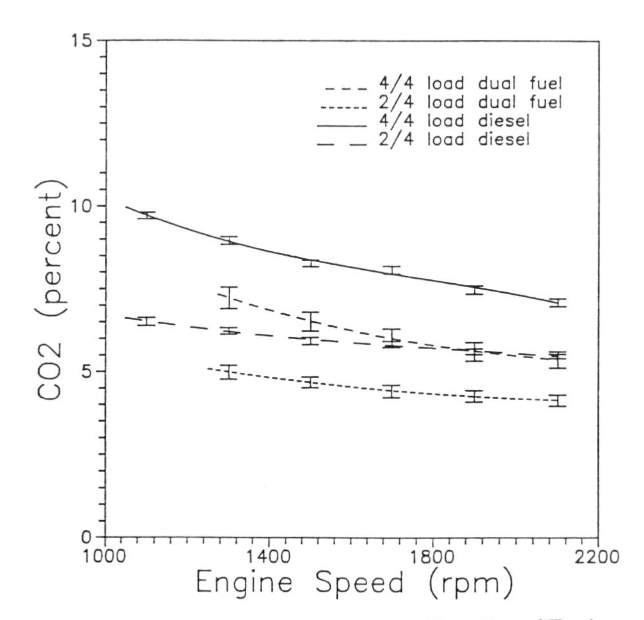

Figure 10 Carbon Dioxide Emission as a Function of Engine Speed for Full and One-Half Loads

ratio is decreased more carbon product, CO_2, is formed compared to hydrogen product, H_2O, thereby increasing CO_2 concentrations for diesel over natural gas fueling. The natural gas used in the study was low in sulfur and thus, its use also significantly reduced SO_2 concentrations as shown in Figure 11. For the cases investigated, the SO_2 emissions were reduced almost linearly with diesel fuel reduction. For full load natural gas, about 15 percent diesel pilot was used and the SO_2 concentration for full load natural gas fueling is about 15 percent the diesel concentration. Because the diesel pilot flowrate per engine stroke was not varied, the SO_2 concentrations for natural gas fueling were near constant.

Particulate emissions from the engine were monitored using a Bosch smoke number method whereby a sample of exhaust was pulled through a filter paper and particulate concentration was indicated using light reflectivity from the sample filter paper. The smoke number plotted as a function of engine speed for full and one-half load conditions for diesel and natural gas fueling are shown in Figure 12. For natural gas fueling, the Bosch smoke numbers are reduced to 1/2 to 1/3 the corresponding diesel only cases. These results indicate a significant reduction in particulate emissions. The reduction of particulate formation is expected since the homogeneity of the air-fuel mixture improves for the natural gas fueling cases. This minimizes the occurance of locally rich zones which can produce particulates. A dilution tunnel has been developed to more accurately monitor particulate emission levels for future reporting.

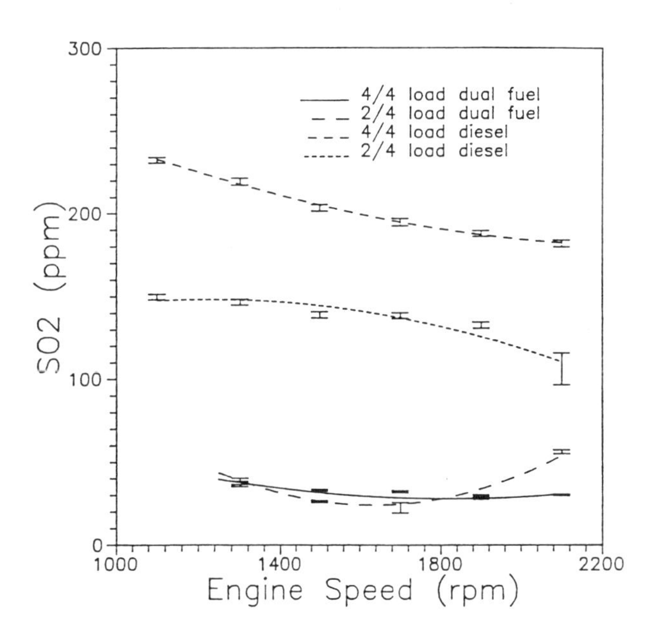

Figure 11 Sulfur Dioxide Emission as a Function of Engine Speed for Full and One-Half Loads

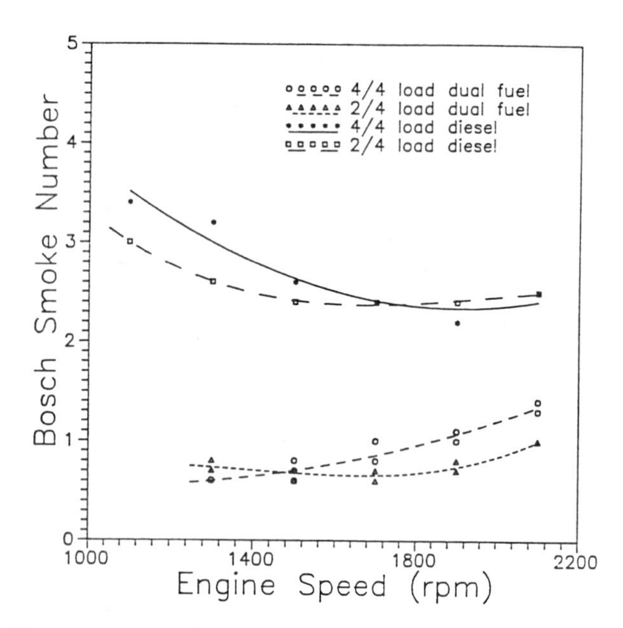

Figure 12 Bosch Smoke Number as a Function of Engine Speed for Full and One-Half Loads

Conclusions

The Caterpillar 3406 engine was operated with natural gas as the primary fuel and diesel as the pilot or ignition fuel. The engine was operated at one-quarter, one-half, three-quarters and full load conditions without encountering knock. The following observations were made for comparing engine operation using natural gas versus diesel:

- For full load operation, the fuel efficiency for natural gas operation was similar to diesel operation,
- For lower loads the fuel efficiency was decreased for natural gas fueling for fixed pilot injection timing,
- NO_x concentrations in the exhaust were reduced at low loads and at high speeds for high loads for natural gas fueling,
- CO emissions were increased for natural gas fueling,
- CO_2 emissions were decreased for natural gas fueling,
- SO_2 concentrations were decreased for natural gas fueling, and
- Particulate emissions were decreased for natural gas fueling.

Other observations which suggest future activities include: the pilot injection quantity and timing is an important parameter for performance and emissions optimization and additional analysis of emissions would be useful to include unburned hydrocarbon measurement and use of a dilution tunnel for particulate measurements.

References

1. Alan P. Gill, "Design Choices for 1990's Low Emission Diesel Engines," SAE paper 880350.
2. R.R. Richards and J.E. Sibley, "Diesel Engine Emissions Control for the 1990's," SAE paper 880346.
3. G. Acker, C.E. Brett, W.J. Schaetzle, and Y.K. Song, "LNG (Liquid Natural Gas) as a Fuel and Refrigerant for Diesel Powered Shrimp Boats," ASME paper 88-ICE-21.
4. Xianhua Ding and Phillip Hill, "Emissions and Fuel Economy of a Prechamber Diesel Engine with Natural Gas Dual-Fueling," SAE paper 860069.
5. T.R. Barbour, M.E. Crouse, and S.S. Lestz, "Gaseous Fuel Utilization in a Light-Duty Diesel Engine," SAE paper 860070.
6. G. Acker, "A Study of the combustion Characteristics of the Dual- Fuel Diesel Engine," Ph.D. Dissertation, The University of Alabama, 1986.
7. Herbert Tesareck, "Investigations Concerning the Employment Possibilities of the Diesel-Gas Process for Reducing Exhaust Emission Especially Soot (Particulate Matters)," SAE paper 750158.
8. J. Boisvert, L.E. Gettel and G.C. Perry, "Particulate Emissions of a Dual-Fuel Caterpillar 3208 Engine," ASME paper 88-ICE-18.
9. P.L. Pitt, "An Ignition System for Ultra Lean Mixtures," Combustion Science and Technology, Vol. 35, pp. 277-285, 1984.
10. Quader, A.A., "Lean Combustion and the Misfire Limit in Spark Ignition Engines," SAE paper 741055, 1974.
11. Anderson, R.W. and Lim, M.T., "Investigation of Misfire in a Fast Burn Spark Ignition Engine," Combus. Sci. and Technol., 1985, Vol 43, pp. 183-196.
12. Heywood, J.B., Internal Combustion Engines, pp. 915. McGraw Hill, New York, 1988.
13. Heywood, J.B., Internal Combustion Engines, pp. 302. McGraw Hill New York, 1988.
14. Obert, E.F., Internal Combustion Engines and Air Pollution, pp. 234-235. Harper and Row Publishers, New York, 1973.
15. Heywood, J.B., Internal Combustion Engines, pp. 372-375, 400-478. McGraw Hill, New York, 1988.

ICE-Vol. 15, Fuels, Controls, and Aftertreatment
For Low Emissions Engines
ASME 1991

EMISSION REDUCTIONS THROUGH PRE-COMBUSTION CHAMBER DESIGN IN A NATURAL GAS, LEAN BURN ENGINE

Michael E. Crane and Steven R. King
Southwest Research Institute
San Antonio, Texas

ABSTRACT

A study was conducted to evaluate the effects of various precombustion chamber design, operating, and control parameters on the exhaust emissions of a natural gas engine. Analysis of the results showed that engine-out total hydrocarbons and oxides of nitrogen (NO_x) can be reduced, relative to conventional methods, through prechamber design. More specifically, a novel staged prechamber yielded significant reductions in NO_x and total hydrocarbon emissions by promoting stable prechamber and main chamber ignition under fuel lean conditions. Precise fuel control was also critical when balancing low emissions and engine efficiency (i.e. fuel economy). The purpose of this paper is to identify and explain positive and deleterious effects of natural gas prechamber design on exhaust emissions.

INTRODUCTION

As natural gas becomes an increasingly attractive alternative to conventional energy sources such as gasoline, diesel fuel, and even electricity, combustion systems designed to exploit the advantages associated with natural gas have risen to the forefront of engine research. One advantage of natural gas is its wide flammability limits, especially to the fuel lean side where ignition of natural gas can occur in the presence of more than 200 percent theoretical air[1]. Operating in this regime translates into improved fuel economy and lower NO_x emissions; however, these improvements are usually at the expense of exhaust HC. The objective of this research was to develop and experimentally verify ways of reducing total HC emissions from prechamber, stratified charge engines operating on natural gas. The effects of these combustion strategies on NO_x were also measured.

A conventional natural gas prechamber is shown in Figure 1. The prechamber usually replaces the diesel fuel injector and seats flush with the firedeck. Communication with the main combustion chamber occurs through a small orifice or throat near the base of the prechamber. A spark plug is located in this area for ignition of the prechamber mixture. During the intake stroke, a very fuel lean, gaseous mixture is introduced into the main chamber. The subsequent compression stroke forces a fraction of the lean main chamber mixture into the smaller volume prechamber where it is supplemented with raw natural gas supplied through the gas inlet valve. The mixture in the prechamber, often near chemically correct, is then ignited via a conventional spark plug and ignition system.

The mixture in the prechamber burns rapidly with a corresponding sharp rise in prechamber pressure. A high energy, burning mixture exits the prechamber through the orifice and initiates combustion of the very lean mixture in the main combustion chamber. This process has the ability to operate under overall leaner conditions than open chamber configurations

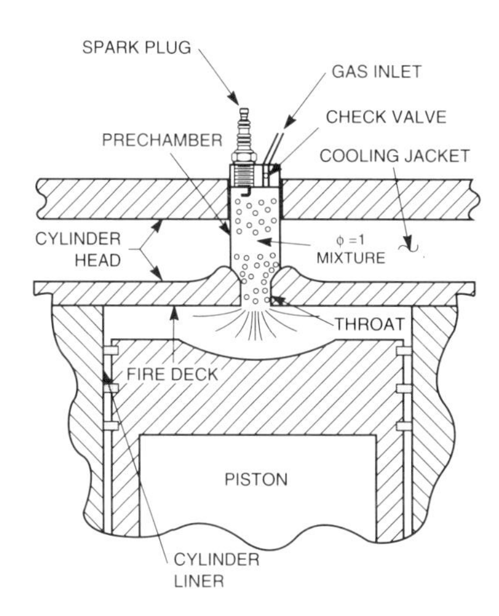

Figure 1. Typical natural gas prechamber installed in a diesel engine cylinder head.

due to the higher ignition energy of the prechamber jet when compared to conventional spark ignition. Ultimately, lower NO_x and CO emissions, less cyclic variability, and reduced combustion duration are realized.

Research was directed at the development and understanding of novel prechamber designs, with the goal of reducing light load/fuel lean HC emissions. This was achieved by understanding what had been investigated in the past, considering those findings in this study's designs, and incorporating novel ideas to address problem areas. As a result, several unique prechambers were designed, fabricated, and characterized on a single cylinder research engine. This report explains the underlying combustion mechanisms resulting in designs that simultaneously reduced HC and NO_x.

EXPERIMENTAL APPARATUS AND PROCEDURES

Experimental Engine Set-Up

A Caterpillar 1Y540 (CAT 1Y540), single cylinder engine was chosen for this investigation. This engine configuration is typically used for oil characterization testing and was well suited to prechamber evaluation due to its inherent size and flexibility. Specific engine features and dimensions are presented in Table 1.

The induction system used the CAT 1Y38 surge tank with a heater mechanism to control inlet air temperature. An exhaust system using piping

Table 1. Engine Features and Dimensions

Bore, mm	137.2
Stroke, mm	165.1
Stock Displacement, L	2.4
Induction	Naturally Aspirated
No. of Cylinders	1
Valve Train	4 vlvs actuated w/pushrods
Compression Ratio	14.5:1 (stock)
Piston Type	Re-entry w/ 3 rings

and an exhaust barrel as specified by the Caterpillar 1Y540 Engine 1-J Test Procedure[2] was also installed. Exhaust backpressure was minimal for all test conditions.

Several modifications were made to the engine prior to testing. A new cylinder head was modified to accept the experimental precombustion chambers in what was formerly the unit injector hole. Modification involved removing the injector and machining a small spot-face on the backside of the firedeck to allow the prechamber base gasket to seat. Furthermore, the injector hole was threaded to provide clamping force on the prechambers.

Additional modifications were required to accept a specially fabricated spark plug near the prechamber throat. A hole was bored that permitted a 195 mm long spark plug to be inserted through the side of the cylinder head and into the prechamber base for throat ignition or flame ionization detection. Water intrusion was sealed against with a high temperature silicone sealer around the outside of the plug's sealing surfaces. Firing pressures were contained with a standard, 14 mm spark plug gasket. This approach worked quite well; there was no evidence of water leakage during any inspections or testing. Final cylinder head modification involved a change to the firedeck to accept a flush mounted, cylinder pressure transducer.

Instrumentation and Cell Set-Up

A summary of the engine test set-up and associated instrumentation is shown schematically in Figure 2. The CAT 1Y540 research engine is shown in the center of the figure. Engine temperature was maintained to

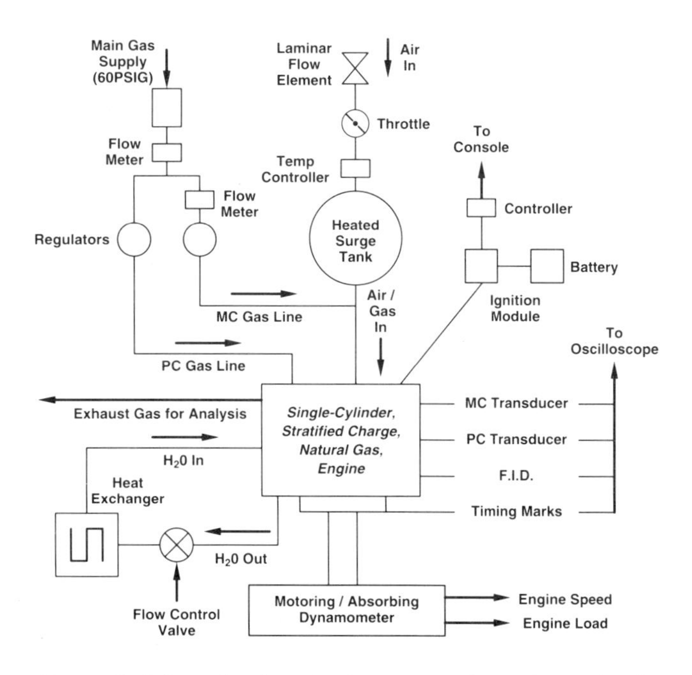

Figure 2. Schematic of engine instrumentation and associated hardware used for evaluation of natural gas prechamber designs.

within 2°C by a shell and tube heat exchanger and was critical since the prechamber was surrounded by coolant.

Natural gas was metered into the engine through regulators providing independent control of flow into the main and precombustion chambers. Main chamber gas was introduced to the intake air supply downstream of the surge tank, approximately 750 mm away from the intake port. Control of gas flow into the prechamber was accomplished by regulating the gas pressure upstream of the prechamber. Gaseous flow rates were adequately measured with positive displacement, Roots-type meters.

Ignition of the inducted fuel and air mixture occurred in the precombustion chamber through either a conventional, 10 mm spark plug gapped to 0.38 mm or the extended length plug gapped to about 0.51 mm (manufacturer's specifications). An Electromotive HPV1 high energy ignition system provided necessary voltage in both cases. A 60-tooth pick-up wheel mounted onto the crankshaft nose served as a shaft encoder for accurate spark timing changes.

Gaseous exhaust emissions including CO, CO_2, total HC, NO_x, and O_2 were measured by a Beckman emissions cart through a heated sample line. Hydrocarbon emissions were measured on a carbon basis (ppm C) and are three times those reported on a straight propane basis.

High speed data encompassed cylinder pressure traces from the precombustion chamber, the main combustion chamber, and a Flame Ionization Detector (FID) signal. Flush mounted and cooled, Kistler 6121A cylinder pressure transducers were used to measure cylinder pressures. Resulting signals were conditioned through Kistler 5004 charge amplifiers and observed through a four channel oscilloscope. Storage of these traces for post test analysis was accomplished with a Norland "Prowler" A-D converter/storage scope.

The final piece of high speed data acquired during this study made use of the aforementioned FID. Supplying a voltage potential across the spark plug not being used for prechamber ignition allowed characterization of the time necessary for the flame to travel a certain distance within the prechamber. This was accomplished by monitoring the time between the ignition event and flame arrival at the FID. The latter event was evident on the oscilloscope as a distinct voltage drop as shown in the upper portion of Figure 3.

Experimental Test Procedure

It was difficult to test all prechambers under identical conditions. Their unique designs expectedly changed their behavior in terms of emissions, lean limit, and combustion stability. However, items that were consistent from test to test are presented in Table 2.

The CAT 1Y540 was broken-in on natural gas with the baseline prechamber configuration. This allowed power and friction to stabilize.

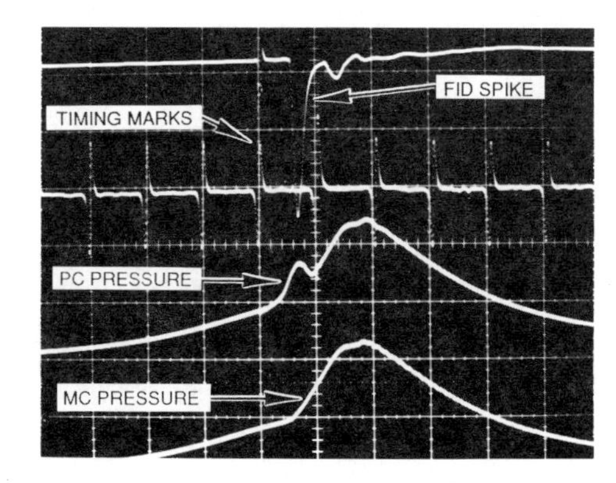

Figure 3. Sample oscilloscope trace showing (from top to bottom): FID pulse, 10 degree increment timing marks (center line is TDC), prechamber pressure trace, and main chamber pressure trace.

Table 2. Standard Test Condition

Oil Sump Temperature, C	72
Coolant Out Temperature, C	85
Engine Speed, rpm	1000
Inlet Air Temperature, C	35
Exhaust Backpressure	Minimal
Inlet Air Humidity, grains	50 - 90
Throttle Position	Wide Open Throttle
PC Fueling Rate	Min, Mid, Max
MC Fueling Rate	That necessary to achieve desired load
Ignition Timing	Sweep
A/F Ratio	~ 30:1
Load	50 psi BMEP

PRECHAMBER DESIGN AND STRATEGY

Design objectives focused on physically incorporating the most successful and state-of-the-art strategies from past investigations into these low emission prechambers. This was accomplished by adopting a modular design that allowed different configurations to be evaluated in a time effective manner without fabricating each prechamber individually.

The first step in the prechamber design process was to determine what aspects of prechamber design would be addressed. It was decided to concentrate on specific design parameters that the literature identified as critical. Parameters that were deemed critical and were accommodated during the design phase included ignition site[3,4], prechamber stoichiometry[5], main chamber stoichiometry[6,7], prechamber volume[3,8,9], orifice size[8,10,11], orifice design[3,4], and fuel mixing[9,12]. Each prechamber design had a role in determining the relative effect of these parameters.

Conceptual approaches were designed to conform to the CAT 1Y540 cylinder head. The cylinder head was sectioned and reproduced on a CAD system to allow for specific prechamber designs. Using the modular approach mentioned earlier, four basic designs were developed to fit the CAT 1Y540 cylinder head and accommodate the necessary instrumentation.

Basic Modular Prechamber

This basic prechamber scheme is shown schematically in Figure 4a and was used in various forms for Tests 1 through 4. The spark plug, fuel inlet valve, and pressure transducer are located in the top part of the prechamber assembly. This interchangeable component is labeled the prechamber cap. The prechamber orifice was fixed in the prechamber base which seated on a modified portion of the cylinder head. The assembly was held together by a prechamber retainer that compressed the cap and base. Threads in the cylinder head secured the retainer and allowed sufficient clamp load.

Each component of the modular design had a specific function. The prechamber base seated securely in the cylinder head and accommodated various orifice sizes and configurations. In addition, it was designed to allow use of a throat-mounted spark plug or an FID. Ignition of the bulk gas or the gases near the throat wall (i.e., Tests 2, 3, and 4) could be achieved with this system. When the top mounted spark plug was used for prechamber ignition, the throat plug was used as the FID. The prechamber cap stacked neatly on the base through an annular groove and no threads. Threadless contacts were desired to prevent seizure and difficult disassembly.

The CAT 1Y540 cylinder head conveniently provided for a wet prechamber since the injector hole communicated with the cylinder head coolant jacket. A wet prechamber helped to prevent detonation, but required that sealing surfaces resulting from the modular design seal against prechamber firing pressures and coolant leaks. The former was accomplished with annealed copper gaskets at every juncture. Each gasket

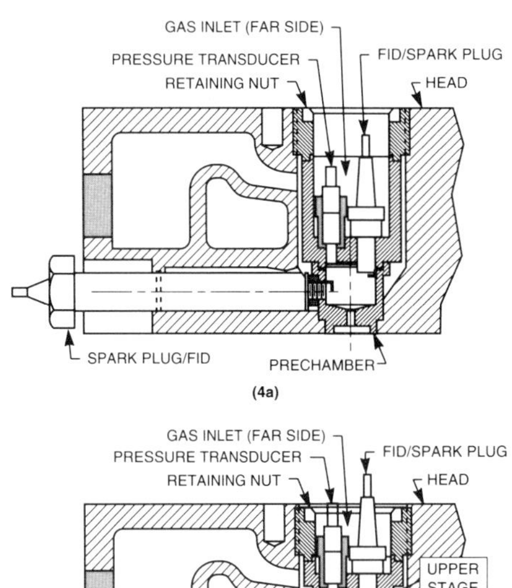

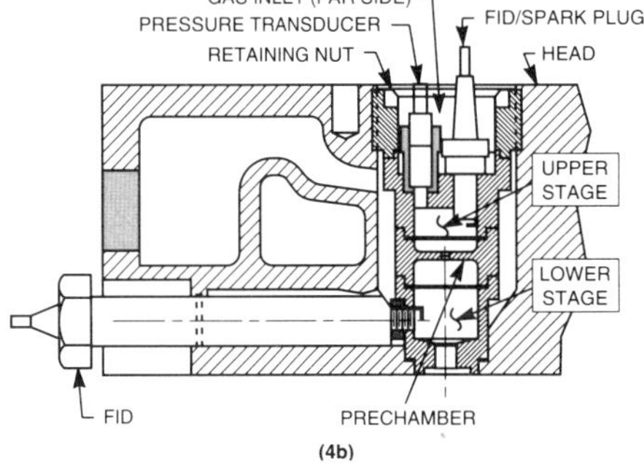

Figure 4. Basic prechambers designs tested for hydrocarbon reductions: (a) conventional design, and (b) Staged Prechamber assembly.

was designed to be approximately 0.08 mm oversized so that the prescribed clamp load would provide adequate crush of each gasket. In this manner, prechamber combustion pressures were sealed from escaping into the cylinder head coolant jacket.

Preventing coolant leaks into the prechamber was crucial since it would have significantly modified characteristic prechamber combustion. These leaks were avoided by simply applying a thin bead of silicon sealer to the circumference of individual mating surfaces. This approach proved to work well.

Combinations of these modular pieces allowed straightforward changes in prechamber volume, orifice diameter, plug location, and prechamber geometry. Design characteristics and test combinations presented in this report are given in Table 3. All prechambers were fabricated from 421 stainless steel.

Staged Prechamber

One of the novel prechambers designed and fabricated for testing was labeled "Staged Prechamber" (Figure 4b). This design features two auxiliary chambers communicating with one another through a secondary orifice. Similar to the previous designs, a premixed fuel and air charge is inducted into the main chamber and subsequently compressed into the Staged Prechamber assembly. Supplemental gas is added to the upper stage of the Staged Prechamber where ignition first occurs through a conventional spark plug.

It was hypothesized that the developed flame kernel in the upper chamber would release a high energy, preliminary torch to ignite the lower stage fuel and air charge. An FID was located near the throat of the lower stage to indicate when the flame front had arrived. Torch ignition in the main combustion chamber was achieved through a primary orifice connecting the main chamber and prechamber assembly.

Two upper stages were designed and fabricated so that the effects of secondary orifice size could be evaluated. Table 3 details the design of the Staged Prechambers (ref. Tests 6, 8, and 9).

Table 3. Prechamber Design Summary

	Prechamber Configuration (by Test Number)							
	1	2	3	4	6	8	9	10
Vol, total pc (ml)	13.27	13.27	13.27	13.27	17.86	17.86	17.86	n/a
Vol, lower pc (ml)	n/a	n/a	n/a	n/a	15.26	15.26	15.26	n/a
Vol, upper pc (ml)	n/a	n/a	n/a	n/a	2.60	2.60	2.60	n/a
% of TCV	6.8	6.8	6.8	6.8	9.0	9.0	9.0	n/a
Dia, prim orifice (mm)	4.75	4.75	9.53	9.53	4.75	9.53	4.75	n/a
Dia, sec orifice (mm)	n/a	n/a	n/a	n/a	9.40	3.18	3.18	n/a
Compression Ratio	13.5	13.5	13.5	13.5	13.1	13.1	13.1	14.4
Ignition Site	TOP	THROAT	THROAT	BULK GAS	TOP	TOP	TOP	CENTRAL MC

EXPERIMENTAL RESULTS

Nine natural gas combustion systems were evaluated as part of this effort:

- Baseline configuration (Test 1)
- Prechamber with throat-mounted spark plug (Test 2)
- Prechamber with throat-mounted spark plug and large orifice (Test 3)
- Prechamber featuring ignition of the bulk gas (Test 4)
- Staged Prechamber with large secondary orifice and small primary orifice (Test 6)
- Staged Prechamber with small secondary orifice and large primary orifice (Test 8)
- Staged Prechamber with small primary and secondary orifices (Test 9)
- Open chamber combustion system (Test 10)

Characteristic HC and NO_x results are briefly summarized for the first seven tests in Figure 5. These curves were generated under the relatively light load of 50 psi BMEP and a minimum HC prechamber fueling rate. Prechamber and total engine air/fuel ratios were held nearly constant during testing; however, as spark timing was increased or prechamber fueling rates were changed, main chamber fueling rate was adjusted accordingly to maintain a constant load. Overall equivalence ratios* for each test configuration are given in Table 4. Equivalence ratio was set to maintain 50 psi bmep and did not vary more than ± 0.025 during any given test.

Results in Table 5 represent a summary of the average emission trends shown in Figure 5. Prechamber designs used in Tests 2, 6, 8, and 9 lowered or maintained total HC emissions relative to the baseline curve. Test 6 demonstrated the greatest improvement by reducing exhaust HC by almost 30 percent. Oxides of nitrogen were reduced relative to the baseline prechamber in Tests 3, 4, 8, and 9. The typical HC-NO_x trade-off is apparent in the first four tests. The latter Staged Prechambers simultaneously lowered exhaust NOx and HC and appear very promising. The only change in fuel consumption was improved fuel economy in Test 2 and significantly worse efficiency in Test 9. Test 9 efficiency decreased due to increased pumping losses across the small diameter orifices in the Staged Prechamber.

Open chamber results are not shown because this configuration would not operate under such lean conditions (average of 0.55 equivalence ratio). Emissions and combustion stability degraded with the open chamber configuration and required that it be run at richer air/fuel ratios and heavier loads at wide open throttle. However, open chamber tests were conducted in an "unoptimized manner". Compression ratio was approximately 10 percent higher than that used in prechamber testing due to reduced clearance volume. Also, the open chamber test was run unthrottled and there was no effort to ensure adequate mixture preparation for the incoming charge. The

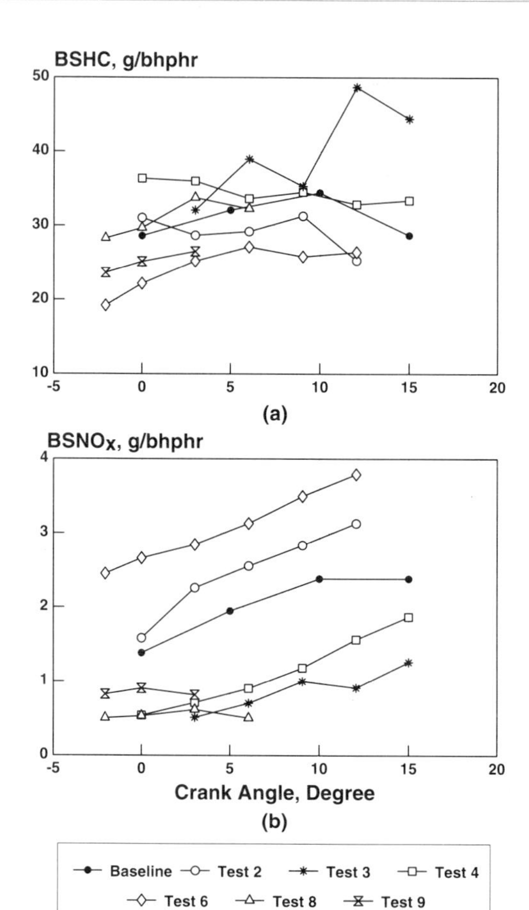

Figure 5. Prechamber design effects on (a) total exhaust HC and (b) oxides of nitrogen. (1000 RPM, WOT, "maximum" prechamber fuel flow rate at 50 psi BMEP)

Table 4. Individual Test A/F Ratios

	Test							
	1	2	3	4	6	8	9	10
Avg. Equiv. Ratio	0.56	0.55	0.52	0.55	0.57	0.55	0.59	N/A
Range	±.015	±.015	±.025	±.005	±.005	±.015	±.005	N/A

Equivalence Ratio = $(A/F)_{stoich} / (A/F)_{actual}$

Table 5. Light Load Emissions Summary
(Results relative to Baseline Test)

Test Number	HC	NOx	Efficiency
Baseline	–	–	–
Test 2	-16%	+25%	+7%
Test 3	+26%	–72%	-2%
Test 4	+13%	–56%	+4%
Test 6	–30%	+50%	–2%
Test 8	+3%	–69%	–4%
Test 9	–14%	–47%	–14%
Open Chamber	n/a	n/a	n/a

spark plug was centrally located between the four valves where the prechamber orifice usually resided.

ANALYSIS OF EXPERIMENTAL RESULTS

Analytical and experimental data are presented to explain mechanisms underlying observed HC, NO_x, and general prechamber performance behavior. Several theories are presented to explain observations related to different prechamber designs.

Hydrocarbon Emissions

Copious amounts of emissions and performance data were taken under various operating conditions during testing. However, HC emissions have been documented to be worst under light load conditions where frequent misfire, poor volumetric efficiency, and/or poor combustion stability prevail. Consequently, the following analyses focus on the relatively light load operation of 50 psi BMEP at 1000 rpm.

Baseline Prechamber. The advantage of operating with a prechamber is evident by realizing that open chamber emissions data are absent from the light load results in Figure 5. Load was controlled solely by main chamber and prechamber fuel metering and it was found that the open chamber configuration was not able to operate under the required, fuel lean, light load conditions. Attempts to do so resulted in excessive amounts of unburned fuel due to frequent misfires. The conventional prechamber served its intended purpose of extending the lean operating limit and thereby reducing NOx. Unfortunately, these reductions in NOx are accompanied by unacceptably high HC emissions.

Test 2 Prechamber. It has been proposed that a strong factor contributing to excessive HC is the location of the spark plug in the prechamber design[3,4,9]. Conventional natural gas prechambers place the spark plug at the top of the prechamber as was the case in the baseline testing. This allows for stable ignition but happens to be the furthest point from the prechamber throat. The result is that unburned prechamber fuel can be forced into the main combustion chamber before it is oxidized.

The prechamber design of Test 2 was identical to the baseline prechamber except that the spark plug was relocated near the prechamber throat. This strategy was successful at reducing HC under light loads by up to 16 percent. Relocating the spark plug near the prechamber throat reduced HC through two controlling mechanisms: reduction of unburned fuel and improved scavenging near the spark plug.

Conventional prechambers force significant fractions of raw prechamber gas into the main combustion chamber as a result of increased pressure ahead of the advancing flame front[4,14]. The result is less fuel energy in the prechamber to generate the desired high energy ignition source for the main chamber. Initiating combustion near the prechamber orifice results in a flame front that travels back into the prechamber–exactly the opposite direction that the baseline prechamber promoted. Unburned fuel is trapped in the high temperature and high pressure, prechamber end gas where it is more completely burned. This is indicated by increased prechamber cylinder pressures (Figure 6). Ultimately, this increase in peak

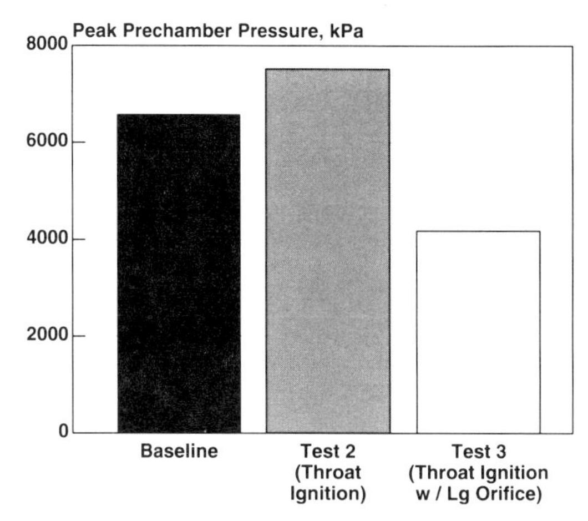

Figure 6. Design effects on peak pressure within the prechamber (5 deg BTDC, 1000 RPM, WOT, Equiv. Ratio = 0.54)

prechamber pressure and rate of pressure rise helps light load operation where HC are worst[4].

The second factor contributing to the observed results in Test 2 is improved, preferential scavenging[3,14]. Locating the spark plug at the top of the prechamber makes it vulnerable to stagnant, burned gases that are difficult to purge during the exhaust event. Lean limit extensions have been reported with throat-mounted plugs that benefit from improved scavenging in the vicinity of the spark plug. This reduces cycle-to-cycle combustion variability by promoting ignition of a fresh air/fuel mixture.

Though prechamber fueling strategy is an operating variable rather than a design variable, its effect was so pronounced that it warrants mention. Test 2 prechamber fueling effects on HC, NO_x, and thermal efficiency are shown in Figure 7. These results are typical for light load operation and demonstrate how prechamber fueling rate must be optimized and tightly controlled. Part (a) shows that approximately 9 to 13 percent of the total fuel charge, by mass, should be admitted to the prechamber for optimal control of exhaust HC. Significantly reduced prechamber fueling rates, such as the 3 percent curve, resulted in too little prechamber energy and erratic engine operation even at advanced ignition timings; hence, the erratic HC and thermal efficiency curves in Figure 7.

Test 3 Prechamber. Brake specific HC and cyclic variability increased with the prechamber design of Test 3. Doubling the prechamber orifice diameter (with respect to Test 2) reduced the prechamber peak pressure by an average 50 percent (Figure 6) and the main chamber peak pressure by 30 percent under light load conditions. The effect was most pronounced by the lack of a characteristic prechamber pressure spike. This is the same result found by previous investigators[8,10]. Increased HC production is the result of reduced in-cylinder turbulence and correspondingly slower burn rates.

These results strongly suggest that turbulence intensity in the main chamber, generated by the kinetic energy of the prechamber jet, can be adjusted by the orifice diameter. A small orifice such as used in Test 2 produces higher in-flow velocities to the main chamber due to the greater pressure differential between prechamber and main chamber. Higher in-flow velocities lead to faster burns caused by greater in-cylinder turbulence. Previous testing with different throat diameters has shown that smaller diameter orifices also provide greater main chamber torch penetration [8,11]

Results also support earlier suggestions that raw prechamber fuel may be forced into the main chamber ahead of the advancing flame front. Recognizing that the prechamber in Test 3 was fueled almost identically to the prechamber in Test 2, the lower cylinder pressure in Test 3 must be the direct result of liberating less fuel energy. This was probably due to raw fuel escaping into the main chamber and out the exhaust before it

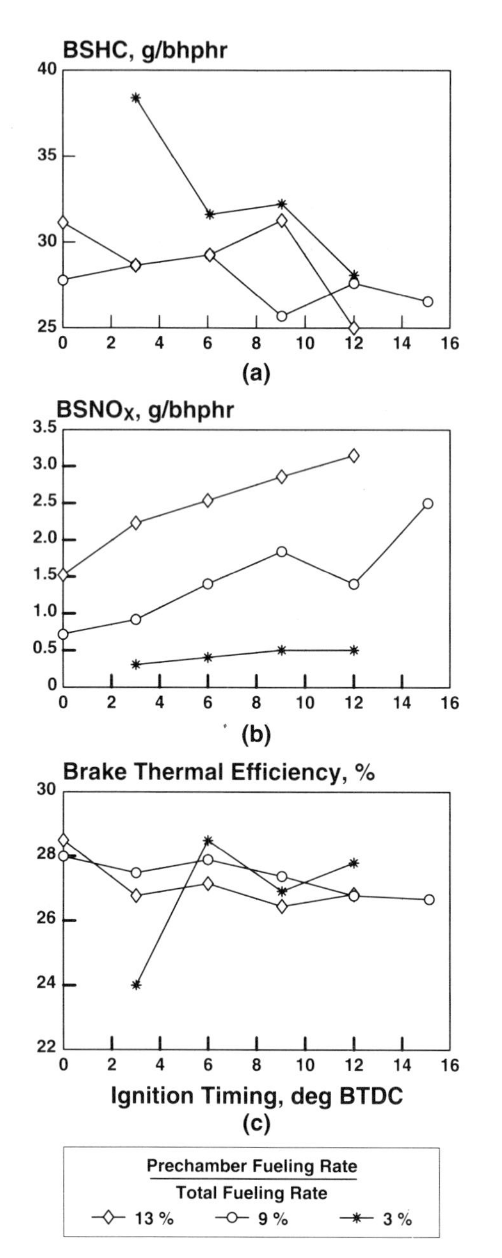

Figure 7. Prechamber fueling rate effects on (a) total exhaust HC, (b) NOx, and (c) brake thermal efficiency (Test #2, 1000 RPM, WOT, 50 psi BMEP).

could be more fully oxidized. This scenario effectively lowers the prechamber's energy producing ability.

If smaller diameter orifices provide greater torch penetration and higher main chamber turbulence which result in reduced HC, then how small is too small for a prechamber orifice? The factor limiting reductions in prechamber orifice diameter is quenching distance of the exiting jet. A flame propagating between two walls (i.e., an orifice) can be extinguished due to high heat transfer if the walls are too close; therefore, it is necessary to identify and relate the effects of pressure, stoichiometry, temperature, and other variables in prechamber design. An excellent, quantitative measure of quenching distance was devised by Adams and used in this study to ensure that orifice diameters were not less than calculated quenching distances[11]. In practice, orifice erosion may also present a practical limit.

Test 4 Prechamber. Prechamber design effects in Test 4 produced higher HC emissions than the baseline prechamber at light load. An extended tip spark plug was installed to initiate combustion in the bulk of the prechamber gas, similar to a centrally located spark plug which reduces combustion duration by allowing the flame kernel to develop radially in all directions rather than unidirectionally within the prechamber.

Cylinder pressure traces showing frequent misfires and high cyclic variability indicated that flame initiation was not readily achieved with this configuration causing excessive HC emissions, possibly due to inhomogeneities within the prechamber. It has been shown[9] that significantly fuel deficient regions exist along the centerline of the prechamber in a natural gas engine. An example of this is shown in Figure 8 where air fuel ratios as lean as 23:1 are observed along the centerline of the prechamber. This problem is further aggravated by high prechamber flow velocities induced through high engine speeds and/or small orifice diameters.

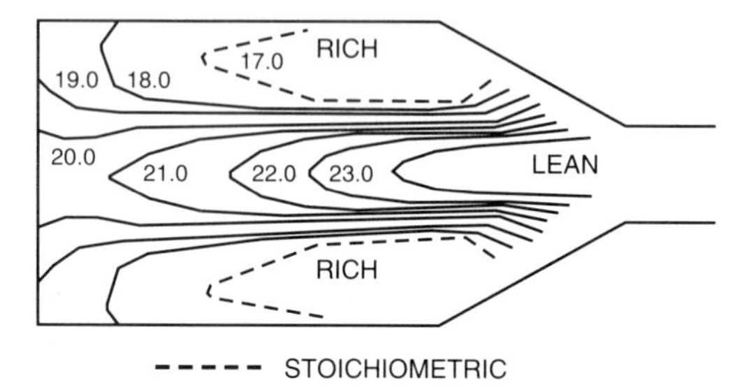

Figure 8. Typical spatial variation of air-fuel ratio in the prechamber of a natural gas engine as predicted by computational fluid dynamics code (12 deg BTDC)[9]

Staged Prechamber Analysis. Unique Staged Prechambers were evaluated in Tests 6, 8, and 9 and showed promise in their ability to reduce HC emissions. Test 6 and Test 9 prechambers reduced HC by approximately 30 percent and 14 percent, respectively, when compared to the baseline configuration. Test 8 maintained the same HC output as the baseline chamber. Differences in NO_x emissions are discussed in the next section.

Though Staged Prechambers differed in secondary and primary orifice diameters, they shared an important characteristic–very fast prechamber burn rates. This is most evident by noticing that the Staged Prechambers are the only chambers run at spark timings *retarded* of TDC (Figure 5). Furthermore, the Staged Prechambers of Tests 6, 8, and 9 reached MBT 3, 4, and 7 degrees sooner than the baseline configuration, respectively.

FID traces (Figure 9a) support the fast prechamber burn hypothesis. Physically, these traces represent the time between prechamber ignition and arrival of the flame front at the prechamber throat. During this time, flame initiation and propagation occur; therefore, these curves provide a time measure of the sum of these two mechanisms.

However, there is another variable that must be accounted for when analyzing these curves. Relative locations of ignition site and FID were not necessarily the same. For instance, the baseline chamber used a top-mounted spark plug for ignition and the throat mounted plug as the FID. The inverse was true in Test 2 although the distance between the two was equivalent. The Staged Prechambers always used the upper stage's spark plug for ignition and the throat plug in the lower stage as the FID. The distance separating these instruments was approximately twice the distance within the conventional type prechambers. To normalize these curves, the distance between the ignition source and FID was divided by the observed prechamber burn duration in Figure 9a. Results (Figure 9b) represent psuedo-average flame speeds for each of the prechambers. The fastest burning prechambers are represented by curves from Tests 9, 6, and 8, respectively.

The Staged Prechamber's ability to burn faster and reduce HC, relative to conventional designs, is further supported by comparing prechamber and main chamber pressure traces. Peak pressure in the upper prechamber stage of Test 6 was almost 950 psi while the peak pressure of the prechamber

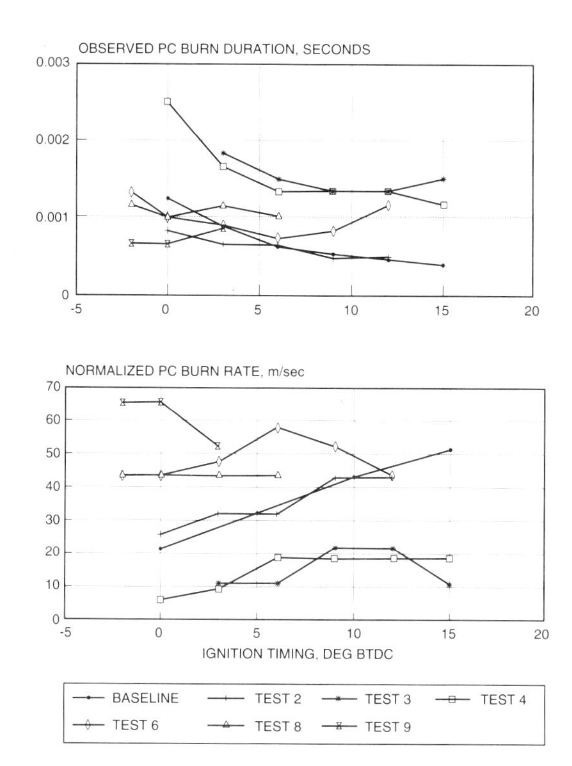

Figure 9. The effect of prechamber design and operating conditions on (a) prechamber burn time, and(b) normalized burn rate

in Test 2 was only 780 psi under similar conditions. This represented an increase of 22 percent. This extra "potential energy" translates into a more energetic ignition system for the lower stage and, ultimately, the main chamber.

Staged Prechamber pressure traces in Figure 10 reveal that secondary and primary orifice diameters significantly affect prechamber and main chamber combustion. Test 8 featured a small diameter secondary (3.2 mm) orifice and large (9.4 mm) diameter primary orifice. The large diameter primary allowed some combustion of the main chamber mixture before peak pressure was reached in the prechamber. Results are similar for Test 6 except that main chamber combustion occured more abruptly. This is evident from the higher rate of pressure rise. In both cases, ignition of the main chamber charge is due to large diameter secondary or primary orifices allowing the upper stage jet to prematurely penetrate into the main chamber. Two small diameter orifices, such as in Test 9, inhibit early penetration of the jet into the main chamber, allowing the burning prechamber mixture to more completely liberate the fuel's chemical energy and supply a stronger ignition source to the main chamber. High energy ignition of the main chamber mixture is evident from the abrupt and highly agitated main chamber pressure trace in Figure 10c. These main chamber pressure fluctuations represent flow reversals between the lower stage and main chamber. Flow reversals were evident in all prechamber tests, but to lesser extents.

Conditions for stable ignition are different between Staged Prechambers as well. Labels in Figure 10 show typical Staged Prechamber "motoring" pressures (pressure prior to ignition). Prechamber and main chamber pressures in Test 6 were almost identical at the time of prechamber ignition. Conversely, main chamber pressures in Tests 8 and 9 were much higher than their respective upper stage pressures. Results show that prechamber pressures and temperatures at the time of prechamber ignition are highest for Test 6, Test 8, and Test 9, respectively. High prechamber pressures (and temperatures) increase flammability limits and make spark ignition easier in the upper stage. The schematic in Figure 11 demonstates that main chamber pressures are highest for Tests 8 and 9 (they are nearly

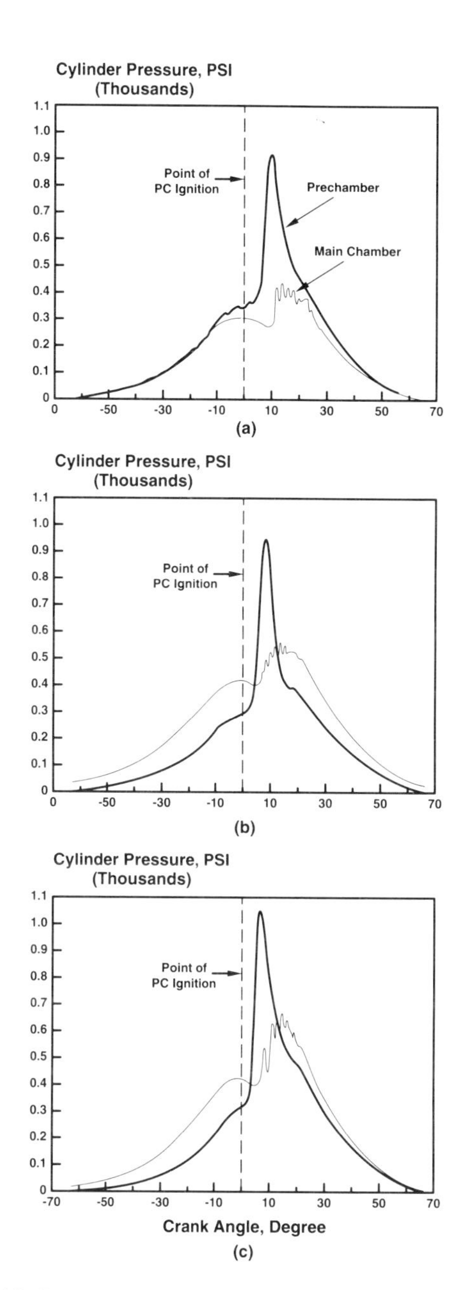

Figure 10. Representative prechamber and main chamber pressure traces for configurations tested in (a) Test 6, (b) Test 8, and (c) Test 9 (0 deg BTDC, 1000 RPM, WOT, "maximum" prechamber fueling rate at 50 psi BMEP)

equal) and lowest for Test 6 at the time of main chamber ignition. Again, flammability limits have been widened by these high pressure conditions which require lower ignition energies (i.e., temperatures) in the main chamber. The advantage in this scenario lies with prechambers from Tests 8 and 9. The air/fuel mixture in the upper prechamber stage is easily ignited without the aid of increased pressure. However, the main chamber mixture is relatively fuel lean and therefore, benefits greatly from the increased pressures that require lower ignition temperatures for flame initiation. Pressure differentials with the small secondary orifices are even greater at more advanced spark timings. Higher main chamber pressures at the time of main chamber ignition could have reduced ignition delay and may explain the higher main chamber pressures in Test 8 (relative to Test 6) for almost equivalent prechamber jet strength (as judged by peak prechamber pressures).

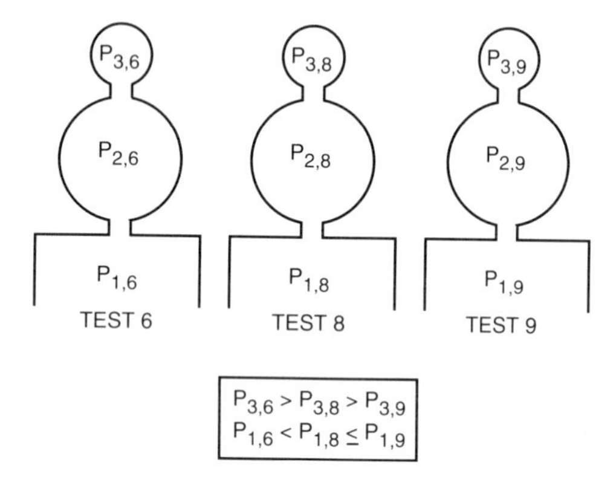

Figure 11. Staged prechamber and main chamber pressure (and temperature) behavior during prechamber and main chamber ignition, respectively

Oxides of Nitrogen Emissions

The previous discussion on HC would not be complete without an explanation of the effects on NO_X emissions. NO_X emissions at the light load test condition were shown earlier in Figure 5. An obvious trend is the dependence of NO_X on ignition timing. Retarded ignition timings drastically cut NO_X emissions in almost every case through the reduction of combustion pressures and temperatures.

Typical NO_X-HC trade-offs are apparent for prechambers tested in Tests 2, 3, 4, and 6: NO_X were reduced at the expense of HC for equivalent ignition timings. An example of the NO_X-HC trade-off under light load operation is shown in Figure 12, where NO_X emissions are plotted against minimum observed HC emissions for a given configuration. Though Test 6 produced the minimum HC, Tests 8 and 9 did the best job of minimizing both NO_X and HC.

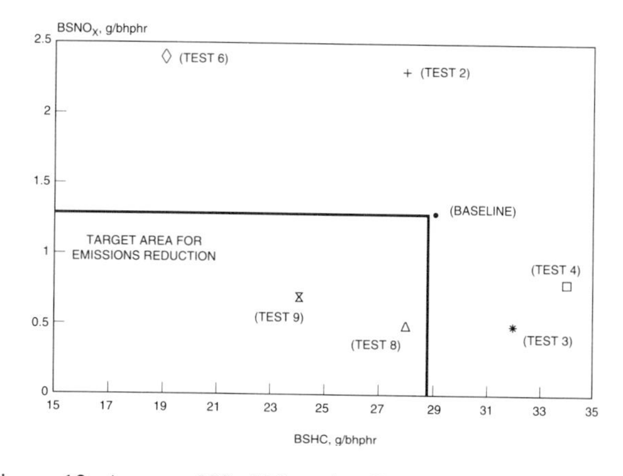

Figure 12. Average NO_X-HC trade-offs for "minimum" HC fueling rate in prechamber (1000 RPM, WOT, 50 psi BMEP)

The promise of the Staged Prechamber approach surfaces when reviewing these HC-NO_X points. Test 8 and 9 prechambers not only reduced or maintained HC emissions relative to the baseline prechamber, but also reduced NO_X by approximately 70 percent and 50 percent, respectively, relative to the baseline prechamber. This may be due to several mechanisms.

Previous research has proposed that the bulk of NO_X emissions are generated primarily in the prechamber and not in the main chamber. Earlier discussion showed that the prechambers generated high pressures and rates of pressure rise, especially when compared to the main chamber. This subjected the relatively small volume of prechamber gases to high temperatures. NO_X generation is governed by the Zeldovich mechanism[15] and is primarily a function of temperature. Thus, the relatively rich prechamber is capable of producing very high pressures and temperatures resulting in the significant production of NO_X emissions. This supports the conclusion that a small prechamber volume is better than a large prechamber volume. Smaller prechambers lower prechamber derived NOx and also minimize thermal losses (by reducing surface-to-volume ratio).

Prechamber fueling rates were just as effective in controlling NO_X production as they were in controlling exhaust HC. Low prechamber fueling rates approaching 3 percent of the total gaseous charge in Test 2 (ref. Fig. 7) minimized NO_X emissions due to partial burns that ultimately lowered prechamber and in-cylinder temperatures.

These curves support the theory that NO_X are prechamber-derived. Reduced prechamber fueling meant enriched main chamber mixtures to maintain the same load. Therefore, the overall air/fuel ratio was held relatively constant (approximately 29:1) while NO_X were observed to drop. Prechamber-derived NO_X drop with fuel lean prechamber mixtures and dominate overall engine-out NO_X production. Similar tendencies were found for the Staged Prechamber in Test 6. Test 8 and 9 results, which featured small diameter (3.2 mm) secondary orifices, contradicted these results. In fact, progressively lower levels of prechamber fueling coupled with higher main chamber fueling rates resulted in progressively higher NO_X under all load conditions.

These latter results suggest that NO_X are not solely prechamber-derived when the secondary orifice is of very small diameter. The small mass of upper stage charge simply is not capable of producing significant amounts of NO_X despite producing very high pressures and temperatures. Rather, it appears that a significant fraction of the engine-out NO_X are created in the lower stage or the main chamber. Analysis of cylinder pressure data suggests that the main combustion chamber is capable of producing significant NO_X when operating with these Staged Prechambers. The peak prechamber pressure for Test 2 (0 deg BTDC) is only 100 psi higher than the main chamber pressure (0 deg BTDC) for Test 9. If prechamber-derived NO_X are significant for the Test 2 prechamber due to high operating pressure/temperature, then it can be inferred that the main combustion chamber in Tests 8 and 9 is capable of producing similarly significant NO_X results. Furthermore, prechamber fueling rate comparisons found that the small secondary orifice chamber tolerated less upper stage fuel than the large secondary orifice of Test 6. To maintain stoichiometry, main chamber fuel was added which further aggravated main chamber-derived NO_X. The role of the lower stage in these processes remains largely undefined.

CONCLUSIONS

1. Prechamber combustion systems can extend the lean operating limit in natural gas engines. This was shown by operating several natural gas prechambers at an average equivalence ratio of 0.55 where a comparable, open chamber design exhibited frequent misfires and unstable torque.
2. Relocating the spark plug to the throat of a conventional prechamber reduced light load HC emissions by 16% but increased NO_X emissions by 25%.
3. Staged Prechambers, incorporating upper and lower stages separated by an orifice, were shown to simultaneously reduce engine-out HC and NO_X emissions up to 14 percent and 47 percent, respectively.
4. Prechamber fueling strategies in conventional or Staged Prechambers must be optimized and tightly controlled to attain desired emissions and efficiencies under varying speeds and loads.
5. Prechamber orifice diameter should be sized to optimize torch penetration, jet strength, main chamber turbulence, and durability.
6. Staged Prechamber designs require more development to confirm the results presented here and to exploit their proposed advantages.

ACKNOWLEDGEMENTS

The author would like to thank W. Liss and A. Wells of the Gas Research Institute for their support of this work and their technical direction; R. Netting (SwRI) whose meticulous technical work made possible the quality results presented in this report; and, A. Brenholtz of Stitt Spark Plugs for supplying the specially designed spark plugs used for throat ignition.

REFERENCES

1. Kanury, A. M., **Introduction to Combustion Phenomena**, Combustion Science and Technology–Volume 2, Gordon and Breach, Science Publishers, Inc., 1975.
2. "Caterpillar 1Y540 Engine 1-J Test Procedure," April 1985.
3. Noguchi, M., S. Sanda, and N. Nakamura, "Development of Toyota Lean Burn Engine," SAE Paper No. 760757.
4. Varde, K. S., M. J. Lubin, "Combustion of a Stratified Charge in a Chamber," IMechE 1974, *The Journal of Automotive Engineering*, pp. 7-10.
5. Beaty, K. D., "Natural Gas Low Emission Engines for Buses; A Scandinavian Project," 2nd International IANGV Conference–Volume II, Buenos Aires, Argentina, October 1990, Paper No. 27.
6. Storrar, A. M., "A Combustion Chamber to Combine High Efficiency, Low Emissions and Wide Fuel Tolerance," *Automotive Engineer*, June/July 1983, pp. 13-15.
7. Personal communications with C. D. Wood on 12/10/90.
8. Snyder, W. E., M. R. Wright, S. G. Dexter, "A Natural Gas Engine Combustion Rig with High-Speed Photography," *Transactions of the ASME-Journal of Engineering for Gas Turbines and Power*, Vol. 110, July 1988, pp. 334-342.
9. Charlton, S. J., D. J. Jager, M. Wilson, A. Shooshtarian, "Computer Modelling and Experimental Investigation of a Lean Burn Natural Gas Engine," SAE Paper No. 900228, International Congress and Exposition, Detroit, MI, February 26-March 2, 1990.
10. GRI Final Report, "Development of Advanced Combustion Technology for Medium and High Speed Natural Gas Engines," Section I, Copy No. 3 of 10, February 1989.
11. Adams, T. G., "Theory and Evaluation of Auxiliary Combustion (Torch) Chambers," ASME Paper No. 780631, Passenger Car Meeting, Troy Hilton, Troy, MI, June 5-9, 1978.
12. Adams, T. G., "Torch Ignition for Combustion Control of Lean Mixtures," SAE Paper No. 790440, Congress and Exposition, Cobo Hall, Detroit, MI, February 26-March 2, 1979.
13. Melton, R. B., Jr., "Minefield Clearance Using Focused Repetitive Shock Waves (U)," SwRI Final Report No. AR-958, September 1974.
14. Personal communication with S. M. Shahed on 1/19/91.
15. Glassman, I., **Combustion**, Academic Press, Inc., 1977.

ICE-Vol. 15, Fuels, Controls, and Aftertreatment
For Low Emissions Engines
ASME 1991

EFFECTS OF SPARK PLUG NUMBER AND LOCATION IN NATURAL GAS ENGINES

Roy C. Meyer, David P. Meyers, and Steven R. King
Engine, Fuel, and Vehicle Research Division
Southwest Research Institute
San Antonio, Texas

William E. Liss
Gas Research Institute
Chicago, Illinois

ABSTRACT

Combustion experiments were conducted on a spark-ignited single-cylinder engine operating on natural gas. A special open chamber cylinder head was designed to accept as many as four (4) spark plugs. Data were obtained to investigate the effects of spark plug quantity and location on NO_x, HC, CO emissions, brake and indicated thermal efficiency, MBT timing, combustion duration, ignition delay, peak cylinder pressure, peak cylinder temperature and heat release over a wide range of equivalence ratios.

INTRODUCTION

Natural gas-fueled engines are widely employed in stationary applications and are gaining acceptance in mobile applications because of emissions reduction potential and alternative energy source concerns. A variety of combustion techniques for achieving low emissions and acceptable performance characteristics can be applied to natural gas engines. Some of the techniques include: stoichiometric burn with three-way catalyst exhaust aftertreatment; lean-burn prechamber; and lean-burn open-chamber with and without oxidation catalyst. This paper seeks to investigate combustion characteristics suitable for a variety of combustion technologies associated with natural gas.

It has long been understood that NO_x production is a function of time, temperature, as well as oxygen and nitrogen concentration.[1] Also well documented are the effects of combustion characteristics.[2] However, although demonstrated on gasoline-fueled engines,[3,4,5] it has not been shown how natural gas combustion, emissions and performance are interrelated by these parameters in terms of ignition source location.

The work conducted in this program looks at the effects of varying spark plug quantity and location on the aforementioned parameters in a natural gas-fueled engine. This program was sponsored by Gas Research Institute (GRI) and conducted at Southwest Research Institute (SwRI).

TECHNICAL APPROACH

In order to achieve the desired objective of the project a single-cylinder engine was chosen as the test engine. A single-cylinder engine was desired in order to alleviate cylinder-to-cylinder combustion variations and coincident data anomalies.

Table 1. Engine Specifications

Engine	LABECCO
No. Cylinders	1
Bore	3.8125 in.
Stroke	3.750 in.
Compression Ratio	10.5:1
Engine Speed	1800 RPM

The engine used in these tests had a bore of 3.8125 in. and a stroke of 3.750 in. Other pertinent engine specifications are shown in Table 1.[6]

The primary objective of this project was to determine the effects of the number of spark plugs and their location on emissions and performance of a natural gas engine. A cylinder head was designed with multiple spark plug locations (see Fig. 1). The head has a flat fire deck which was used with a flat-top piston, giving a disk-shaped combustion chamber.

Fig. 1. Cylinder Head

A single centrally-located plug (configuration 'A') is typical of many diesel engine conversions where the fuel injector is directly replaced with a spark plug. Multiple ignition sources and faster burn rates were anticipated by using all four plugs (configuration 'B') while slowest burn rates were expected by using a single plug located as far from the center as possible (configuration 'C'). A legend as to plug location in reference to the cylinder head is shown in Fig. 2.

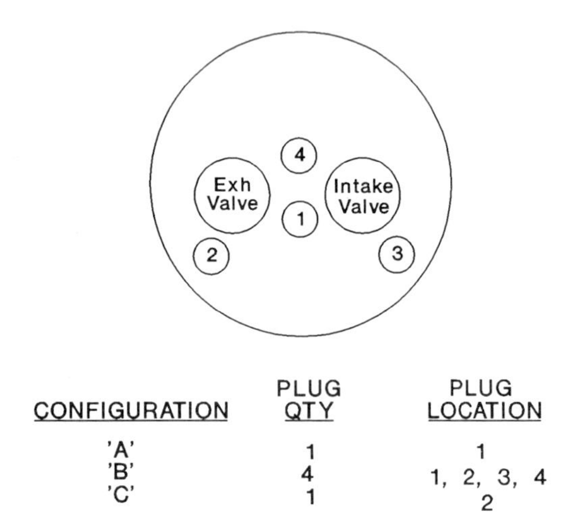

Fig. 2. Spark Plug Location Legend

Equivalence ratio sweeps from 1.00 to 0.65 were performed for each of the three plug configurations at 1800 RPM. For each point in the equivalence ratio sweep, MBT (minimum best torque) spark timing was determined and data were taken.

The reader should bear in mind that brake-specific emissions and efficiency data contained in this report at lean equivalence ratios are not representative of values obtainable with optimized, turbocharged lean-burn natural gas engines due to low brake power.

EXPERIMENTAL SET-UP

Monitored variables included fuel, coolant and lubricant temperatures, pressures at several points, intake air temperature, and exhaust temperature at the exhaust port. The computer data collection and analysis system was programmed to calculate brake horsepower (BHP), brake and indicated thermal efficiency, IMEP, in-cylinder gas temperature, ignition delay, combustion duration, heat release and standard deviation of peak cylinder pressure.

The fuel mass flow rate, engine rpm, fluid and component temperatures and pressures were recorded by a Hewlett-Packard 3497 low-speed data acquisition system which is tied to an HP 1000 Series E minicomputer. Each of the engine parameters was recorded by the HP 3497 once every 15 seconds. Ten of these samples were averaged for each of the test runs.

High speed data were obtained for each test point using a cylinder pressure transducer (Kistler model 6121) and a shaft encoder. This data was based on a hundred cycle average for each half degree of crank angle. All indicated data were obtained by use of this high speed data acquisition system. IMEP, heat release, gas temperature, and combustion duration were determined using Heywood's "Fully Mixed Model".[7]

The natural gas fuel composition changed with time in SwRI's laboratory to the extent that corrections to equivalence ratio, BTU content, and fuel flow rate were made on all data using gas composition measured by an on-line gas chromatograph. It was observed that the methane concentration varied from 92 percent to 96 percent (mole fraction).

RESULTS

Performance Results

Fig. 3 shows the performance results of the three different spark plug configurations tested. It can be seen that for maximum brake horsepower and IMEP, the multiple 'B' plug configuration is better than the centrally-located 'A' configuration which in turn is better than the peripheral 'C' configuration.

Analysis of the ITE and BTE shows that while the centrally-located 'A' plug configuration and the multiple 'B' plug configuration are considerably better that the peripheral 'C' configuration, the multiple 'B' configuration is actually no better than the centrally- located 'A' plug configuration rich of 0.75 equivalence ratio. Below 0.75 equivalence ratio the multiple plug configuration, 'B', is able to extend the lean misfire limit, presumably due to the multiple ignition sources which yield a higher probability of initiating combustion.

It should be noted that all thermal efficiencies depicted in Fig. 3 are based on the higher heating value of the fuel. This was done in accordance with the natural gas utility industry standard of basing everything on higher heating value.

Emission Results

Fig. 4 shows the emissions results taken for the three configurations tested. Examination of Fig. 4 reveals that NO_x production for the centrally-located 'A' and multiple 'B' configurations is very similar. The combustion duration and peak cylinder temperatures discussed in the next figure are also very similar for the centrally-located 'A' and multiple 'B' configurations. NO_x production for the peripheral 'C' configuration is significantly lower and can be attributed to the fact that, while the combustion duration is longer, the peak cylinder temperature is not as high. This lower NO_x result illustrates the pronounced sensitivity of NO_x production to temperature rather than time since this "slow-burn" peripheral 'C' configuration exhibits much longer burning duration.

Examination of hydrocarbon emissions reveals an interesting condition. The different plug configurations yield almost inverse results whether rich or lean of approximately 0.80 equivalence ratio. Above 0.80 equivalence ratio, the peripheral 'C' configuration produces the least hydrocarbons while below 0.80 equivalence ratio this configuration produces the most hydrocarbon emissions. The cause of this phenomena is again associated with the combustion duration. This may be thought of in the following way: there is a certain quantity of fuel-air mixture trapped in the crevice volume above the top ring land but below the piston crown surface. The quicker burn times associated with the multiple plugs result in the main chamber combustion being completed early while there is still a considerable mass of fuel-air mixture caught in the crevice volume. Conversely, the peripheral 'C' configuration with its longer burn time allows more time for the crevice mixture to escape, thus enabling it to be burned with the main charge. Below 0.80 equivalence ratio this slow-burning mechanism detrimentally effects hydrocarbon emissions. For this case the duration of burn at the peripheral 'C' plug configuration is sufficiently long that the end of combustion is not complete prior to the exhaust valve opening and flame quenching. However, the centrally-located 'A' and multiple 'B' plug configurations with their relatively fast burn rates are able to complete combustion before the exhaust valve is opened.

The CO_2 and O_2 trends verify the accuracy of the equivalence ratio determination. The fact that the data points fall consistently on top of one another and in a straight line indicates that the changes in fuel composition have been dealt with accurately.

As for CO emissions, we see that the peripheral 'C' plug configuration exhibits the highest CO. This is a result of the relatively poor combustion characteristics of this configuration. Again we see that the centrally-located 'A' configuration exhibits characteristics very similar to that of the multiple 'B' plug configuration.

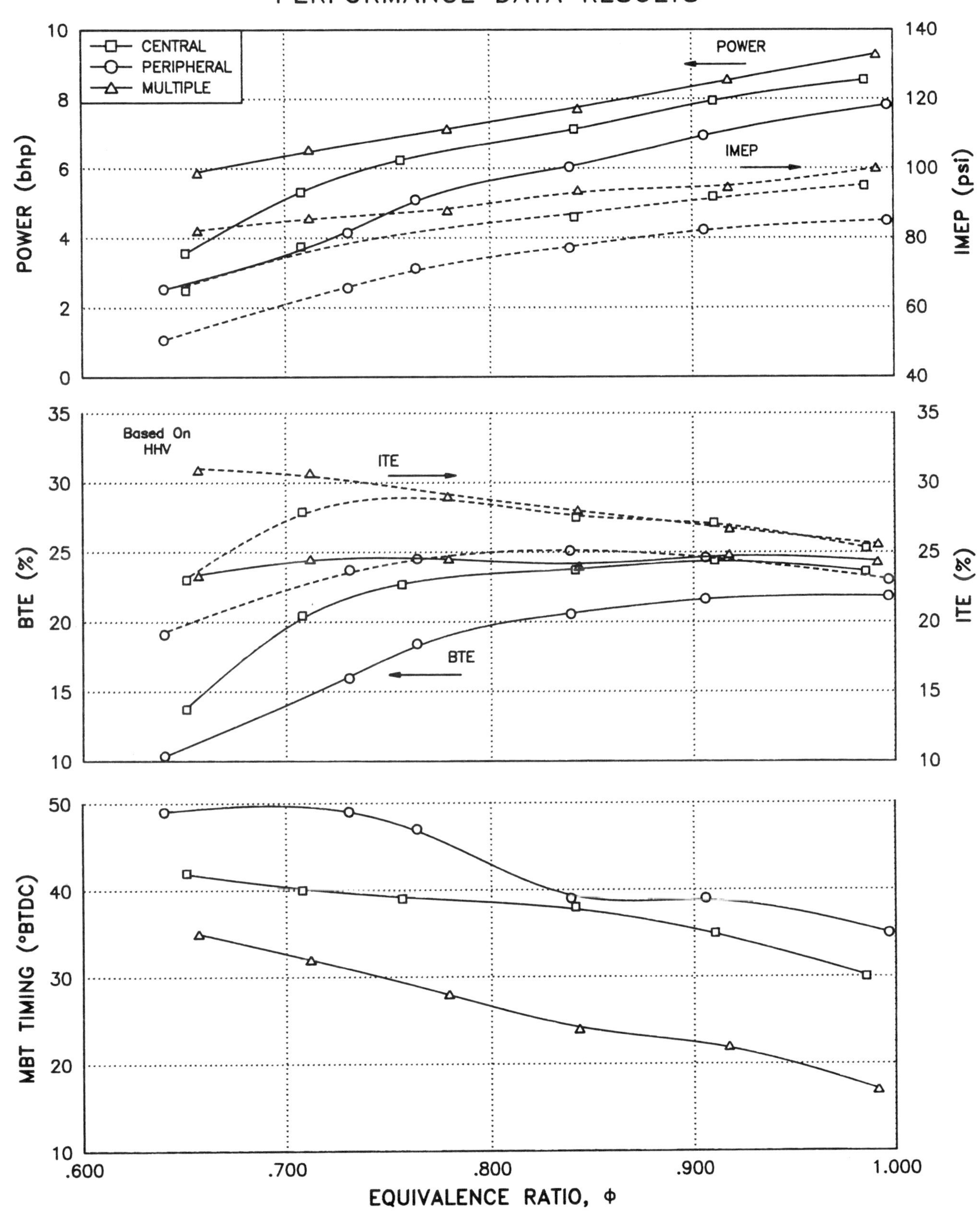

Fig. 3. Performance Results

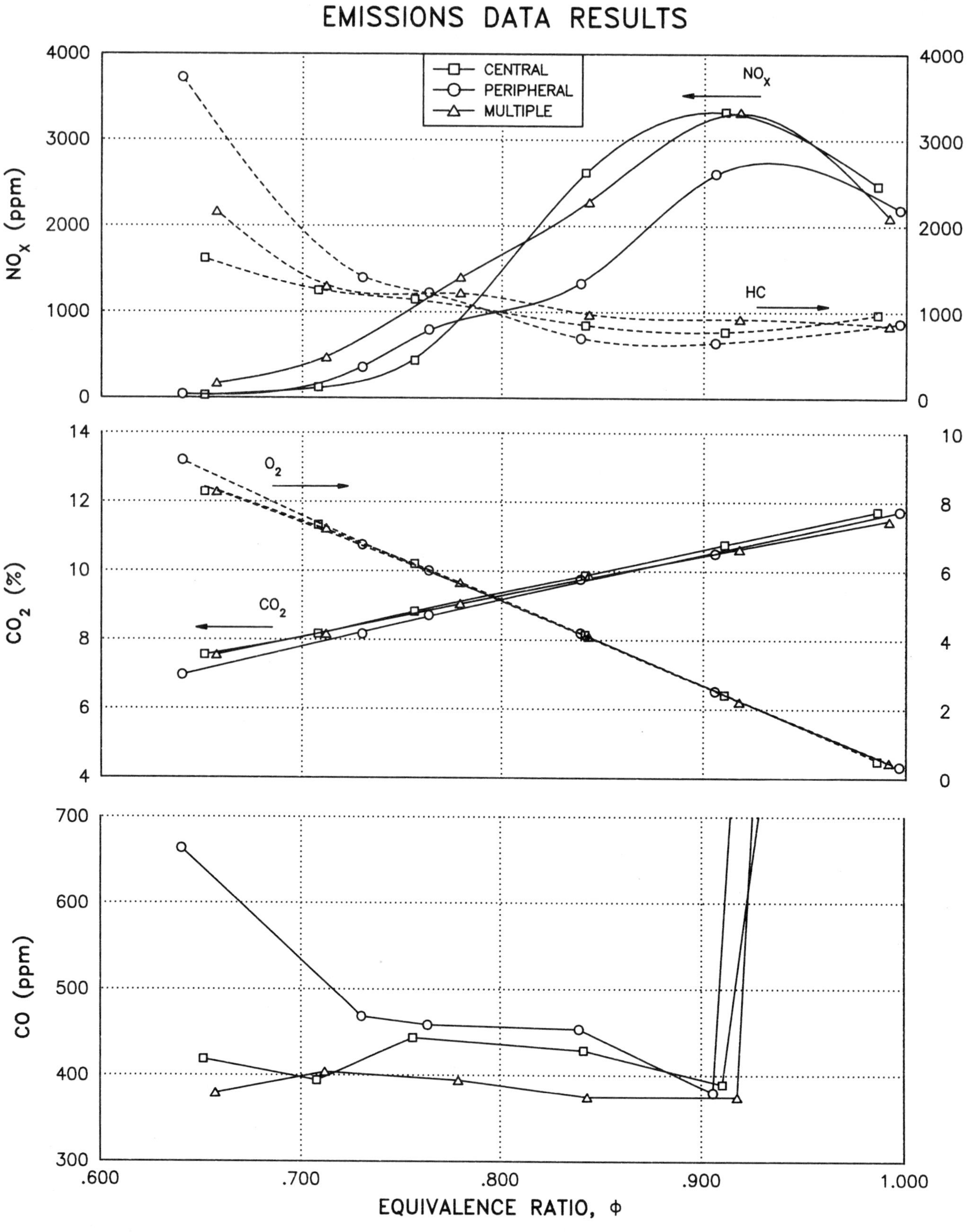

EMISSIONS DATA RESULTS

Fig. 4. Emissions Results

All emissions data shown in Fig. 4 are shown on a volume fraction basis. On a brake specific basis (Fig. 5), the centrally-located 'A' plug configuration shows the most benefits in the extremely lean region for NO_X production. The multiple plug configuration shows the highest specific NO_X production and can be attributed to peak cylinder temperature as is explained in the next section. As expected, the multiple plug configuration has the lowest specific HC production and is, as stated earlier, due to its lower incidence of misfire.

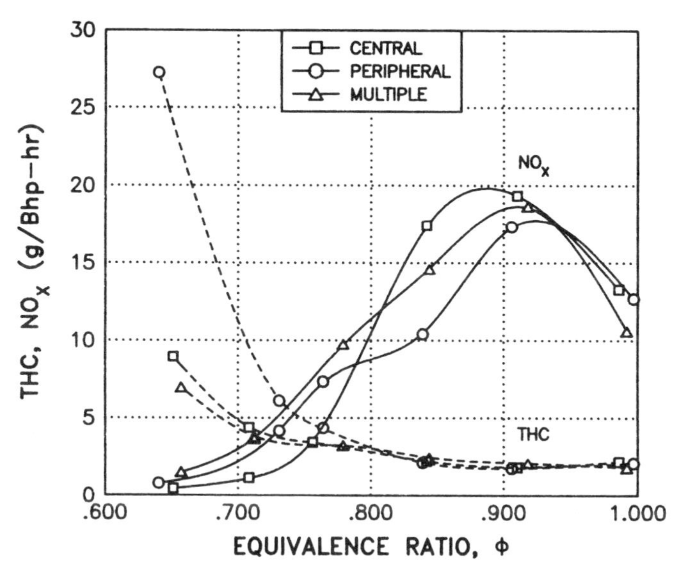

Fig. 5. Brake Specific Emissions Results

Combustion Analysis Results

Fig. 6 is a summary of the results of the combustion analysis. The peak cylinder pressures for the multiple 'B' plug configuration are consistently higher than either the centrally-located 'A' or peripheral 'C' configurations. This is due to the faster burn rate associated with the multiple plug configuration which has the peak pressure occurring closer to top-dead-center and therefore high peak cylinder temperatures. Since at the leaner equivalence ratios the multiple plug configuration is still firing more consistently and closer to TDC, the average peak cylinder pressure is higher.

Analysis of peak mass-averaged cylinder temperatures shows that the multiple plug configuration produces higher temperatures than the other configurations. This was caused by the relatively short combustion durations near TDC. Conversely, the centrally-located 'A' and peripheral 'C' configurations have considerably lower temperatures due to longer combustion durations occurring well after TDC.

Analysis of the cumulative heat release results reveals that the multiple 'B' plug configuration actually released slightly less heat than the centrally-located 'A' configuration. This is believed to be due to the fact that, although almost identical amounts of total heat were released, the higher peak in-cylinder gas temperatures and the earlier occurrence of combustion for the multiple plug configuration allowed more heat to be transferred through the combustion surfaces, therefore, the apparent heat release was lower. The relatively low heat release associated with the peripheral 'C' plug location is due to the relatively slow combustion and lower gas temperatures of that configuration.

The remaining graphs of Fig. 6 depict ignition delay (0-10 percent cumulative heat release) and combustion duration (10-90 percent cumulative heat release) for the three configurations. Interestingly, at equivalence ratios greater than 0.75 the centrally-located 'A' plug configuration burns as fast as the multiple 'B' plug configuration. Yet, in terms of ignition delay, the centrally-located 'A' configuration is comparable to the peripheral 'C' configuration. The peripheral 'C' configuration produces significantly longer combustion durations when compared to the other two configurations. This is believed to be the main cause of the performance and emissions effects presented above.

CONCLUSION

In conclusion we see that the multiple plug configuration 'B' has the largest advantage in performance, in that it produces greater horsepower and higher thermal efficiencies for all equivalence ratios. It also has lower HC production at the leaner equivalence ratios due to its reduced misfire rate. However, its "ultra fast" burn rates result in higher NO_X production than the other configurations at equally lean equivalence ratios. The peripheral 'C' plug location exhibited both poor performance and high emissions production over most of the range except between 0.80 and 1.00 equivalence ratio where it actually produced slightly lower NO_X and hydrocarbon emissions. The centrally-located 'A' plug configuration offers the best compromise in that it has adequate performance and low emissions production when compared to the other configurations.

Finally, if the different plug configurations are associated with different combustion rates it is evident that there is an optimal burn rate for each equivalence ratio. This optimal burn rate must be slow enough to minimize NO_X production and yet fast enough to produce adequate performance characteristics. It appears fortunate for those converting existing diesel engines to natural gas that the single centrally-located spark plug produces burn rates near this optimum.

ACKNOWLEDGEMENTS

The authors of this paper would like to thank the following people whose efforts have made this paper possible:

Allen Wells,	GRI
Chip Wood,	SwRI
Shannon Vinyard,	SwRI
Steven King,	SwRI
John Hedrick,	SwRI
Elvan Sekula,	SwRI
Dennis Kneifel,	SwRI
David Dowell,	SwRI
Janie Gonzalez,	SwRI

REFERENCES

1. Lavoie, G. and Heywood J., 1970, Experimental and Theoretical Study of Nitric Oxide Formation in Internal Combustion Engines, *Combustion Science and Technology*, Vol 1, 313-326.
2. Fleming, R. and O'Neal, G., 1985 Potential for Improving the Efficiency of a Spark Ignition Engine for Natural Gas Fuel, paper presented at International Fuels and Lubricants Meeting, SAE paper 852073.
3. Thring, R., 1979, The Effects of Varying Combustion Rate in Spark-Ignited Engines, SAE paper 790387.
4. Johnson, J. H., 1962, Effect of Swirl on Flame Propagation in a Spark-Ignition Engine, SAE paper 565C.
5. Sakai, Y. and Miyazaki, H. 1973, Effect of Combustion Chamber Shape on Nitrogen Oxides, JSAE paper no. 6 1974.
6. Ainsley, W. and Cleveland, A., 1955, CLR Oil Test Engine, paper presented at SAE Fuels and Lubricants Meeting, paper #641.
7. Heywood, J., Internal Combustion Engine Fundamentals, New York, New York, McGraw-Hill Publishing Company, 1988.

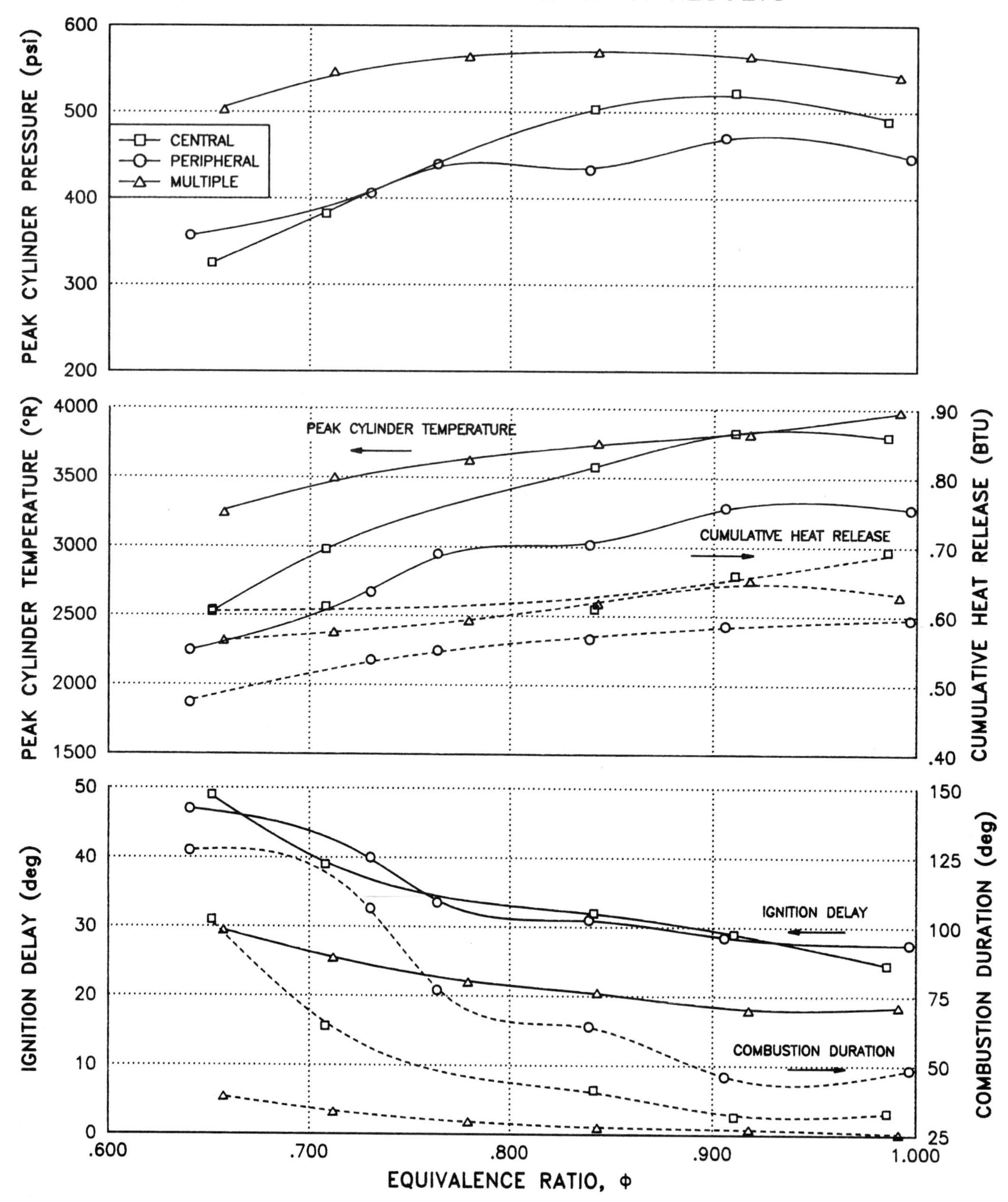

Fig. 6. Combustion Analysis Results

ICE-Vol. 15, Fuels, Controls, and Aftertreatment
For Low Emissions Engines
ASME 1991

DEVELOPMENT OF THE COOPER-BESSEMER CLEANBURN™ GAS-DIESEL (DUAL-FUEL) ENGINE

Donald T. Blizzard
Cooper-Bessemer Reciprocating Products Division
Cooper Industries
Grove City, Pennsylvania

Frederick S. Schaub and Jesse G. Smith
Cooper-Bessemer Reciprocating Products Division
Cooper Industries
Mount Vernon, Ohio

ABSTRACT

NOx emission legislation requirements for large bore internal combustion engines have required engine manufacturers to continue to develop and improve techniques for exhaust emission reduction. This paper describes the development of the Cooper-Bessemer CleanBurn™ gas-diesel (dual-fuel) engine that results in NOx reductions of up to 92% as compared with an uncontrolled gas-diesel engine. Historically, the gas-diesel and diesel engine combustion systems have not responded to similar techniques of NOx reduction that have been successful on straight spark-ignited natural gas burning engines. NOx levels of a nominal 1.0 gm/BHP-HR, equal to the spark-ignited natural gas fueled engine, have been achieved for the gas-diesel and are described. In addition, the higher opacity exhaust plume which is characteristic of the gas-diesel combustion is significantly reduced or eliminated. This achievement is considered to be a major breakthrough, and the concept can be applied to both new and retrofit applications.

NOMENCLATURE

- AMB - Ambient Temperature, °F
- AMP - Air Manifold Pressure, Inches Mercury Absolute
- BMEP - Brake Mean Effective Pressure, PSI
- BSFC - Brake Specific Fuel Consumption (in BTU/BHP-HR)
- BTC - Before Top Center
- CO - Carbon Monoxide, gm/BHP-HR
- DEV - Deviation, Firing Pressure, PSI
- FP - Firing Pressure, PSIG
- HCt - Total Hydrocarbons, gm/BHP-HR
- IGN - Ignition Timing, Crank Degrees Before Top Center
- KW - Kilowatt
- NOx - Oxides of Nitrogen, gm/BHP-HR
- NBN - Normal Butane Number, a Measure of Fuel Sensitivity
- NOP - Nozzle Opening Pressure
- PC - Port Closure, Fuel Pump, in Crank Degrees Before Top Center
- PO - Pilot Oil
- PSIG - Pounds per Square Inch Gauge
- PTT - Pre Turbine Temperature, °F
- RPM - Revolutions per Minute
- Y - Yellow

BACKGROUND

The following is a brief background regarding Cooper-Bessemer's history of exhaust emission control programs. The CleanBurn™ low NOx level natural gas burning engine was introduced in 1977 for integral engine compressor units serving the gas industry. This low NOx integral two-stroke gas engine uses a jet cell chamber operating on the fuel-rich side of stoichiometric. The main chamber operates with very fuel-lean combustion by virtue of high scavenging air supercharging level. The net result of the diverse combustion is to reduce the NOx from a range of 15-20 gm/BHP-HR to 1-3 gm/BHP-HR. Over 165 integral gas engine compressors have been sold by Cooper-Bessemer since 1977. Because of this technology, virtually every engine of this type sold today is of the CleanBurn™ configuration.

The four-cycle gas-diesel (dual-fuel) engine is a popular driver for generators in municipal, sewage treatment and base-load cogeneration power plants because of its durability and because of its ability to burn 94% natural gas as the most economical fuel, yet change over to full diesel should the gaseous fuel supply be interrupted for any reason. Operating in the dual-fuel mode, approximately 6% fuel oil is injected and compression ignited which in turn ignites the remaining 94% gaseous mixture. By contrast to the straight gas engine, equivalent low emissions technology for large slow speed four-cycle dual-fuel engines had never developed.

Best Available Control Technology (BACT) levels were achieved by certain parametric adjustments including retarding injection timing of the approximately 6% pilot fuel oil. BACT NOx levels were typically 5 gm/BHP-HR and gas-diesel opacity levels were within the mandated limit of 20%.

However, even though opacity levels met all requirements, the inherent yellowish colored exhaust plume exhibited during dual-fuel operation presented a psychological problem with customers and the general public feeling that anything visible is bad and not acceptable. Complaints received regarding these legal but visible emissions became one of the major driving forces behind our R&D Program.

This paper contains relevant detail regarding the type engine being marketed with the new combustion system, the development engine and facility, the extensive effort set forth to the gas-diesel engine in prior years using more conventional reduction techniques, and the significance to the dual-fuel CleanBurn™ program of having earlier applied the two-cycle spark-fired integral engine reduction system to the four-cycle power engine.

THE LSB-6 LABORATORY ENGINE

The operating data that follows, unless noted, was generated using a six-cylinder in-line 15-1/2 inch bore 22 inch stroke engine. The model series is LSB-6 and it is fitted with top end parts identical to the LSVB series. The LSVB series has enjoyed a comparatively high new engine production in the past several years particularly in the cogeneration market. See Figure 1.

Figure 1
Cogeneration Installation of Cooper-Bessemer
LSVB Dual-Fuel Engines Delivering 8386 BHP Per Engine

The LSB-6 was installed in the Mount Vernon, Ohio Research and Development Laboratory in mid 1988 and has been fitted for quick conversion to spark-gas, diesel, or gas-diesel modes. The laboratory setup provides precise recording of data by one-quarter crank degree encoding and series 900 Hewlett Packard computer equipment provides display and hard copy pressure-time and rate-of-heat-release data for each test condition. Fuel selection is pipeline quality natural gas, propane, and methanol. Blending equipment is on-site for normal butane to study sensitivity (Reference 1), as well as a twenty-ton source of CO_2 to simulate digester gas. Figures 2 and 3 illustrate the Laboratory installation of the LSB-6 and identify supporting systems. Figures 4 and 5 are sectional drawings showing general engine design of the LSVB and LSB.

The six vertical cylinders of the Laboratory LSB-6 engine allow quick, cost controlled changes to combustion components, and hydraulic shrink cams permit a timing change for one group of cams that can be accomplished in two hours.

Correlation of data between the Mount Vernon Laboratory six-cylinder unit and the production sixteen- and twenty-cylinder units has been excellent regarding turbocharger matching pressure and air mass flows, emission levels, and general performance. The slightly higher friction of the proportionally larger bottom end of the LSB-6 (LSVB-12 crankshaft) requires a 1.5% downward fuel correction to project the V angle production units. This correlation not only provides credible projection of new technology to production units, but also allows accurate and inexpensive handling of Marketing quotes to address non-standard ratings, fuels, and emission requirements.

GAS-DIESEL OPEN CHAMBER ACTIVITY

Preliminary Work

Extensive effort was pursued with gas-diesel open chamber emission control techniques which have been of interest in prior years since the birth of the Clean Air Act. Reference 2 includes data recorded twenty years ago using readily available NOx reduction methods including ignition timing retard, reduced air manifold temperature, increased air rate, and torque reduction. More recent work has included tests using a four valve 13 inch bore, 16 inch stroke single cylinder laboratory engine with a combustion chamber geometry and fuel admission system, similar to that of the LSVB. Tests were completed exploring reduced pilot oil, various nozzle configurations, very high pilot oil injection rate, high injection pressure, and cylinder swirl. High injection rate

Figure 2
Laboratory Installation
of LSB-6 Power Engine

Figure 3
Data Acquisition and Control Room
for Laboratory LSB-6 Power Engine

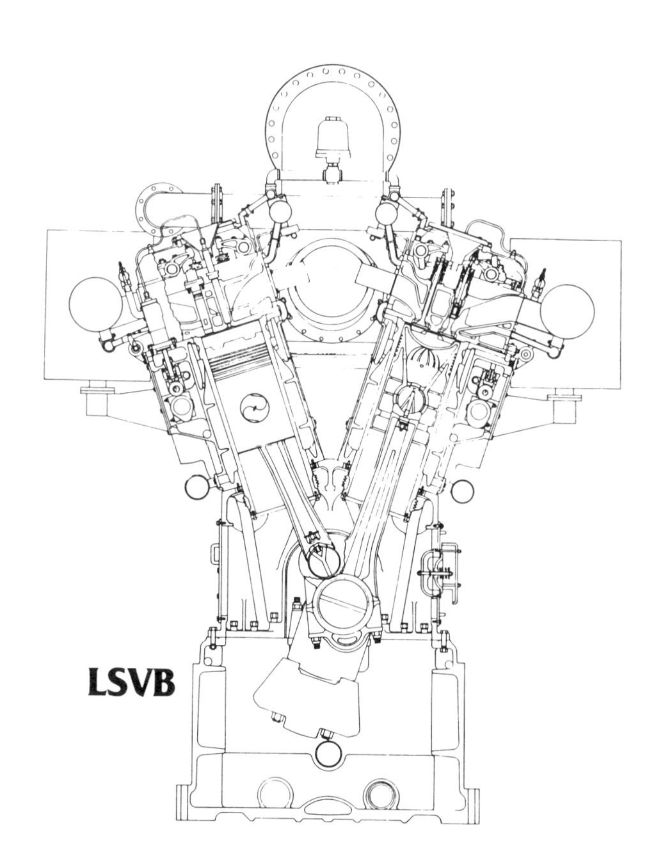

Figure 4
Sectional Drawing of the LSVB Engine

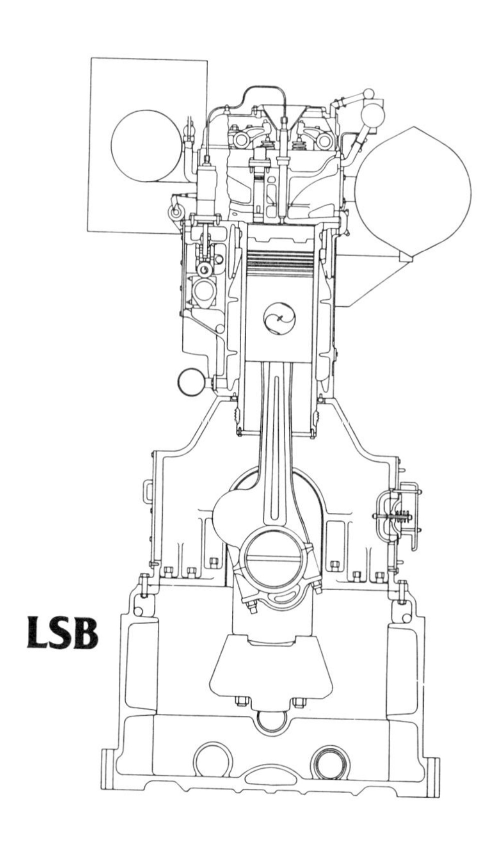

Figure 5
Sectional Drawing of the LSB Engine

was provided by utilizing a non-standard injection pump with a 50% increase in plunger area. High injection pressure was accomplished by back pressuring the injection nozzle and raising the pressure from a conventional 7,000 PSIG to double that level. Swirl was provided by inlet valve shrouding which is described later in this paper.

Work Using the Laboratory LSB-6

The most promising of these tests were repeated using the LSB-6 Laboratory engine. Figure 6, lines 1 through 5 illustrates for the standard gas-diesel engine the single effect of controlling air manifold pressure through a range limited by detonation on the low pressure end and by high fuel consumption or combustion roughness on the high end. Note that NOx can be reduced by increasing the air manifold pressure from 55 to 72 inches mercury absolute pressure, however, NOx levels below 4 gm/BHP-HR could not be achieved, and a fuel consumption penalty of 5% accompanies the reduced NOx. Also accompanying the reduced NOx, at higher air manifold pressure, is a rise in unburned hydrocarbons and a rise in carbon monoxide (not shown). An incremental improvement in exhaust opacity is noted with increased air manifold pressure.

Line 6 of Figure 6 shows the effect of reducing the pilot oil quantity to the standard gas-diesel nozzle while using line 4 air manifold pressure. NOx is reduced an additional 20% compared to line 4, however, engine combustion regularity becomes limiting as the peak pressure standard deviation exceeds seventy.

Lines 7 through 12 were run with 4 hole nozzles and illustrate the benefit of fuel nozzles available to those installations not requiring instant changeover to full diesel mode. The 4 hole nozzle does not allow adequate fuel flow to operate 100% diesel. All data was recorded with "line 4" air manifold pressure. Line 7 compared to line 4 shows a drop in NOx for like quantities of pilot oil and a slight improvement in fuel consumption. The line 8 reduction in pilot oil to 2.8% drops the NOx to 2.2 gm/BHP-HR, but engine combustion regularity is unacceptable at a standard firing pressure deviation of 78 psi.

Lines 9 and 10 using larger .021" diameter 4 hole fuel nozzles show negligible change in numbers attributed to the nozzle configuration change, while lines 11 and 12 using .018" diameter 4 hole fuel nozzles and 18° injection timing, compared to lines 7 and 8 show a negative fuel and NOx benefit associated with further injection retard to 15° BTC.

Line 13 shows the negligible effect of increasing the nozzle opening pressure from 3,500 PSI to 5,500 PSI in the interest of improving penetration. The data shows little change from lines 11 and 12.

Other test work performed using the standard open chamber gas-diesel configuration included variations of main valve timing to effect an improvement in either fuel mixing and/or fuel sweep (waste) during the valve overlap period. No reduction in fuel consumption, opacity, or NOx was found as compared to the standard timing. (See discussion section of this paper for additional detail regarding role of unburned hydrocarbons.)

Figure 7 provides a summary of activity directed to the study of in-cylinder swirl. Figure 8 is a photograph of a standard inlet valve with a 115° shroud welded to the valve head. Configuration A of Figure 7 illustrates the shroud position used to generate a clockwise in-cylinder mixture motion as viewing downward in the cylinder. Rate of

LSB-6 80°F AMBIENT
Gas-Diesel Mode - 200 BMEP

Line	Detail	% PO	AMP	NOx	HCt	BSFC	FP	DEV	Smoke % Opacity	PTT
1	Std Noz 18° PC	5.7%	55	11.5	1.0	6231	1301	49	20Y	1207
2	Std Noz 18° PC	5.6%	59	8.2	1.8	6254	1249	52	20Y	1158
3	Std Noz 18° PC	5.7%	63	5.2	3.3	6352	1207	49	15Y	1095
4	Std Noz 18° PC	6.8%	67	4.4	4.9	6528	1196	48	15Y	1054
5	Std Noz 18° PC	5.5%	72	4.1	7.3	6592	1203	40	10Y	1014
6	Reduced PO	3.8%	67	3.2	4.6	6530	1165	71	15Y	1072
7	4 Hole .018, 18° PC	6.0%	67	3.2	6.1	6396	1168	55	15Y	1049
8	4 Hole .018, 18° PC	2.8%	67	2.2	4.3	6510	1110	78	15Y	1088
9	4 Hole .021, 18° PC	5.6%	67	3.9	4.8	6526	1180	51	10Y	1066
10	4 Hole .021, 18° PC	4.3%	67	3.2	4.7	6519	1141	53	10Y	1077
11	4 Hole .018, 15° PC	5.9%	67	3.2	4.6	6530	1126	69	10Y	1082
12	4 Hole .018, 15° PC	4.2%	67	2.8	4.6	6576	1090	69	10Y	1103
13	Raise NOP 5500 PSIG	5.5%	67	3.6	4.0	6618	1129	63	10Y	1103

Figure 6
Summary of Most Promising Open Chamber Data

combustion was increased significantly, requiring resetting of the pilot oil pump port closure from 18° BTC to 9° BTC to limit peak pressures at a 1,200 PSIG level.

Comparison of Figure 7 configuration A data to Figure 6 line 4 data shows that high swirl improved BSFC level slightly but actually increased NOx. Note that for the same approximate air manifold pressure, pre-turbine temperature has been increased, indicating the degree of flow restriction imposed by the shroud to the normal free flow, four valve configuration. Engine air flow data substantiates this observation. Unburned hydrocarbon levels are lower for Figure 7 configuration A, however, we attribute this to exhaust oxidation at the higher pre-turbine temperature rather than to improved fuel air mixing or other factors accountable to the shrouds.

Configurations B, C, and D are included to show results of some of the other shroud positions evaluated. Configuration B, compared to Configuration A, produced slightly higher peak pressures, lower BSFC, and higher NOx, with increased combustion roughness. Configurations C and D show reduced peak pressures compared to Configurations A or B, and show comparable NOx and total unburned hydrocarbon levels to Configuration A.

Figure 9 shows a plot of NOx versus BSFC for the open chamber gas-diesel data lines of Figures 6 and 7. The general trend is a sweep to the lower right hand corner of the grid showing a conventional tradeoff of increased fuel consumption for beneficial reductions in NOx. The target region for combined NOx and BSFC improvement is of course the lower left hand corner of the grid. Lines 6 through 12, working with reduced pilot, tend to move in the desired direction but meet the barrier of marginal combustion as indicated by peak pressure standard deviation numbers greater than 50. Lines A through D of Figure 7 are clearly on the undesirable side of the standard engine variable air rate response curve shown by lines 2 through 5 of Figure 6.

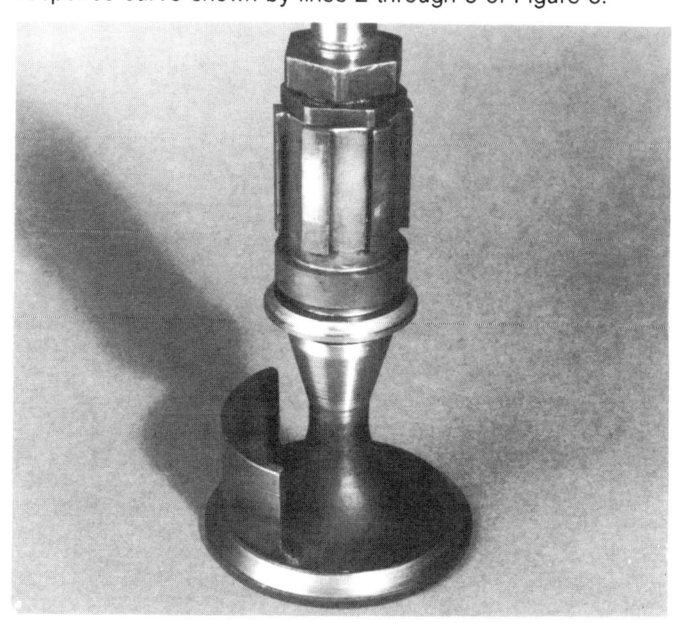

Figure 8
Inlet Valve with 115° Shroud

LSVB-6-80 AMB
Gas-Diesel Mode - 200 BMEP
Shrouded Inlet Valves

Configuration "A"
Fuel Nozzle 9° PC

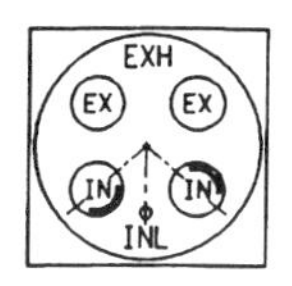

%PO	AMP	NOx	HCt	BSFC	FP
6.3	65	5.5	1.7	6455	1196

DEV	SMOKE % OPACITY	PTT
58	10Y	1130

Configuration "B"
Fuel Nozzle 9° PC

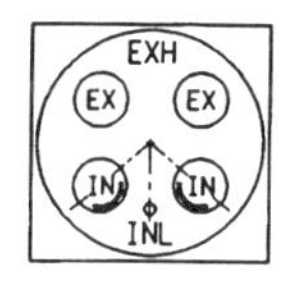

%PO	AMP	NOx	HCt	BSFC	FP
6.6	65	7.5	1.8	6382	1216

DEV	SMOKE % OPACITY	PTT
71	10Y	1141

Configuration "C"
Fuel Nozzle 9° PC

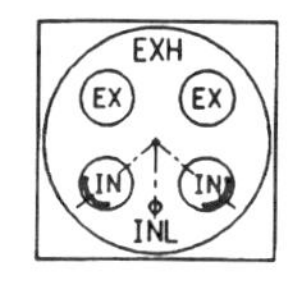

%PO	AMP	NOx	HCt	BSFC	FP
6.1	65	5.4	1.6	6530	1088

DEV	SMOKE % OPACITY	PTT
52	15Y	1186

Configuration "D"
Fuel Nozzle 9° PC

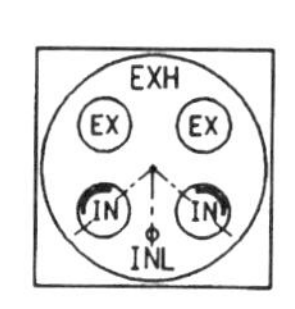

%PO	AMP	NOx	HCt	BSFC	FP
6.7	65	5.4	1.6	6443	1106

DEV	SMOKE % OPACITY	PTT
58	10Y	1179

Figure 7
Summary of Work Investigating
In-Cylinder Swirl

APPLICATION OF THE SPARK-FIRED TORCH CELL TO THE FOUR-CYCLE ENGINE

Commercial application of the two-cycle integral compressor engine torch cell concept to the LSV four-cycle power engine line was introduced in 1985 to Southern Energy Company at Elba Island, Georgia in the interest of achieving acceptable power quality for a spark-ignited gas engine serving very tight frequency requirements. A later phase of this program included turbocharger modifications resulting in a full

CleanBurn™ package achieving not only precise power but also control of oxides of nitrogen to levels in the 2.0 gm/BHP-HR range. More recent development has extended this NOx level to the 1 gm/BHP-HR level and has resulted in sale to date of fourteen units with ratings up to 180 BMEP and for fuels ranging from LNG to digester gas.

Fuel Air Ratio and NOx Formation

Figure 10 has appeared in many technical presentations (Reference 3) on the subject of NOx control for combustion processes and shows NOx formation as a function of fuel-air ratio. It provides the basic arguments for understanding the CleanBurn™ NOx control mechanism. Essentially the cylinder main chamber is controlled to be very fuel lean and combustion temperatures are thereby controlled to levels sufficient to limit NOx formation. The pilot or torch chamber is fuel rich and NOx is thereby limited by the available oxygen. The main chamber is represented by the right side of Figure 10, and the torch chamber is represented by the left.

Design and Performance Details of the Spark-Fired CleanBurn™ LSVB Engine

Figure 11 illustrates the spark-fired torch cell installation to the LSVB engine line. Cell geometry and volume are basically unchanged from that of the two-cycle integral engine line but nozzle direction is tailored to accommodate the more vertical access available for the LSVB engine line. As with the larger bore two-cycle engines, two cells per cylinder are used which net combustion smoothness and BSFC/NOx gains compared to one cell per cylinder. The spark-plug and pilot gas check valve system are identical to that of the two-cycle.

Figure 12 tabulates performance numbers for variable air manifold pressure at two spark advances and Figure 13 highlights the NOx versus BSFC relationships against those of the open chamber gas-diesel shown earlier as Figure 9. There is a marked reduction in NOx levels and opacity levels for the spark-fired CleanBurn™ system. These data were generated on the same engine, using the same fuel gas admission system and the same combustion chamber. Only the method of ignition is different. The obvious dilemma was to account for the higher NOx level that occurs, when only 3 to 6 percent of the total fuel used is number 2 fuel oil and the remainder is natural gas as compared with 100% natural gas.

THE PILOT OIL TORCH CELL

The Development Program

After exploring every practical avenue of open-chamber pilot oil injection options for the gas-diesel engine, it became apparent that diffusion burning of the pilot oil charge could not be tolerated. Pilot oil penetration and pattern variations did little to control NOx. Cylinder swirl dramatically affected burning rate but not NOx formation. Reduced pilot oil quantities either with the standard full diesel capability nozzles or with reduced area "gas-diesel only" nozzles produced unacceptable combustion quality and fuel consumption. It appeared that no matter how we handled the pilot injection, there was a significant quantity of pilot oil and associated main

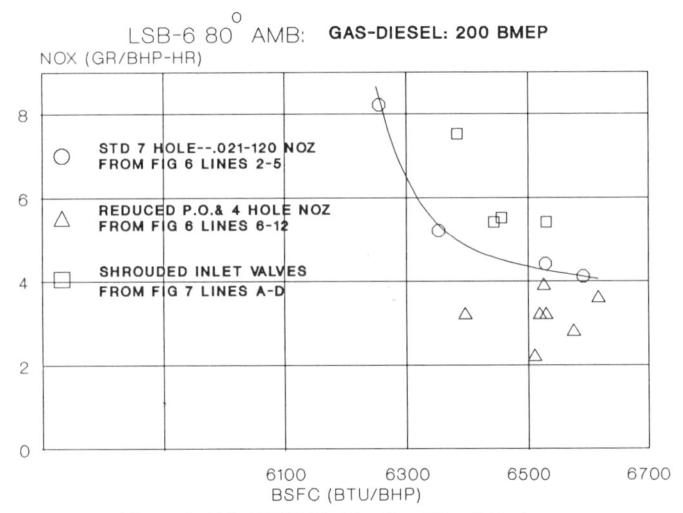

Figure 9 NOx/BSFC Plot for Gas-Diesel Mode

chamber fuel gas being burned in a manner contributing substantially to NOx formation.

Experience and success with the spark-fired CleanBurn™ torch cell system, directed gas-diesel program effort away from open chamber diffusion burning. The desired benefit of the pilot oil torch cell was to not only control NOx by rich burning, but also to hopefully minimize the quantity of pilot oil involved since its role as an ignition energy source would apply only to the cell rather than the main chamber.

Initial pilot oil torch cell work was assigned to the 13 inch bore, 16 inch stroke single-cylinder laboratory engine. In the interest of rugged design and low maintenance, a pintle type nozzle was suggested and a matrix of tests completed exploring cell size and geometry, cell nozzle diameter and

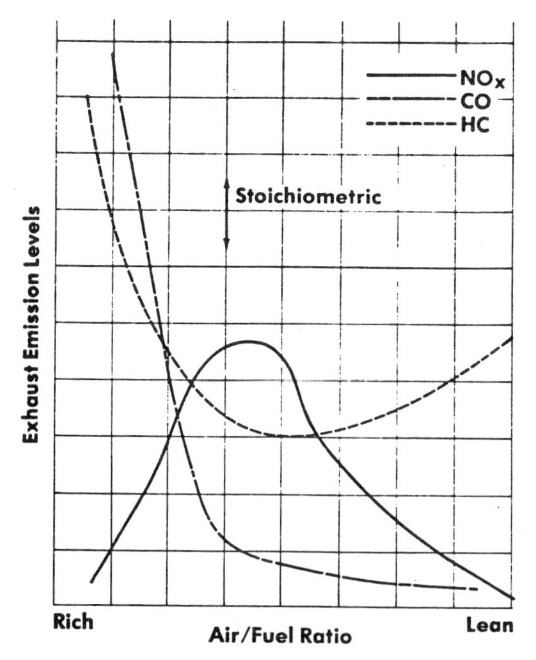

Figure 10
Effect of Air/Fuel Ratio
On Exhaust Emissions Internal Combustion Engines

LSB-6 80°F AMBIENT
Spark-Gas CleanBurn™ Mode 180 & 200 BMEP

Line	Detail	AMP	NOx	HCt	BSFC	FP	DEV	Smoke % Opacity	PTT
14	5° Ign 180 BMEP	60	1.5	4.2	6319	1040	41	< 5%	1077
15	5° Ign 180 BMEP	62	1.0	4.7	6400	1045	42	< 5%	1049
16	8° Ign 180 BMEP	64	1.0	6.3	6326	1177	29	< 5%	991
17	4° Ign 200 BMEP	75	1.0	5.4	6240	1206	51	< 5%	1020
18	4° Ign 200 BMEP	75	1.2	5.0	6250	1200	50	< 5%	1040

Figure 12
Summary of Spark Gas CleanBurn™ Data

length/diameter ratio, cell direction, and main chamber geometry. It was gratifying to find that, although these tests netted some gain in performance, the basic system displayed wide margin for operation and did not require precise control of operating parameters.

The program transition from the single-cylinder engine to the laboratory LSB-6 was easily accomplished. A series of tests were performed to establish final cell volume turbocharger matching pilot fuel quantity and timing, load acceptance, and fuel gas sensitivity (NBN) limits for an array of test parameters. Flame-out conditions were defined as a function of load and minimum air manifold temperatures.

System Description

Figure 14 illustrates the pilot oil fired torch cell installation in the LSVB cylinder head. The pilot oil injector is a modified commercial truck pintle nozzle. Typical pintle performance includes some blossom and ignition of the initial injection followed by the main penetrating injection which intentionally strikes the uncooled nozzle portion of the torch cell. Cell charge motion as viewed in Figure 14 is clockwise on the engine compression stroke as fuel lean main chamber mixture enters the cell, and counterclockwise during torch cell combustion. Figure 15 illustrates the cell and injector, and Figure 16 shows the actual installation in the cylinder head. Pilot oil injection pumps are modified units chain driven from the engine camshaft.

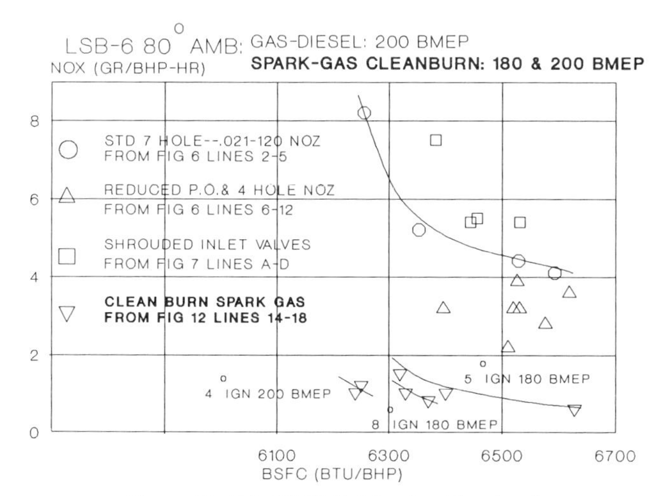

Figure 13 NOx/BSFC Plot for Gas-Diesel and Spark-Gas CleanBurn™ Modes

Performance Results

Figure 17 tabulates performance data as a function of air manifold pressure for 20° BTC and for 23° pilot oil pump port closure. Comparison to Figure 12 for the spark-gas CleanBurn™ mode shows excellent correlation. Figure 18 highlights the gas-diesel CleanBurn™ mode NOx versus BSFC relationships against those of the open chamber gas-diesel and spark-fired CleanBurn™ engine.

Figure 19 makes a comparison between the present day uncontrolled gas-diesel engine and the CleanBurn™ gas-diesel engine. NOx has been reduced 92%, opacity reduced from an objectionable yellow level to virtually a clear stack, and pre turbine temperatures reduced over 200°. The jet cell allows a reduction of pilot oil from a nominal 6% to near 1%, yet retains an excellent ignition source for the fuel gas. The dual-fuel capability of this engine is still retained by this concept with full diesel rating being achieved by changeover to the conventional fuel system. During the gas-diesel mode the main fuel racks are set to zero delivery.

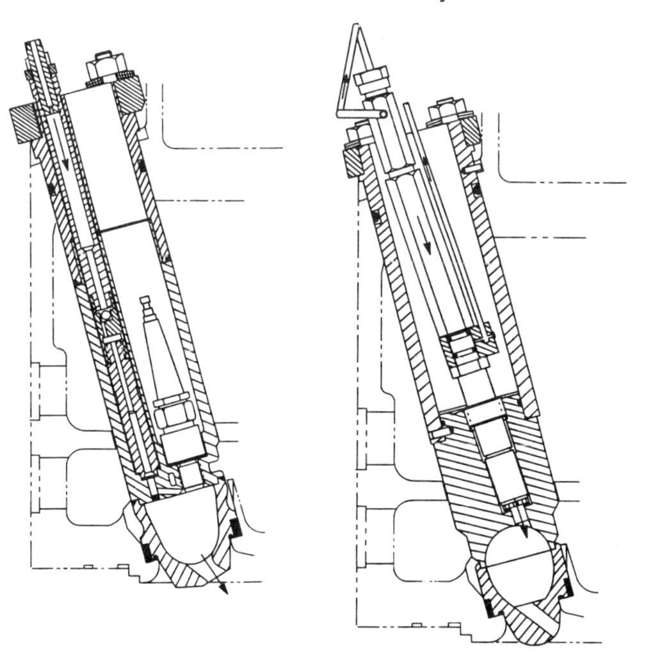

Figure 11
Torch Cell for
Spark-Gas Power Engine

Figure 14
Torch Cell for
Gas-Diesel Engine

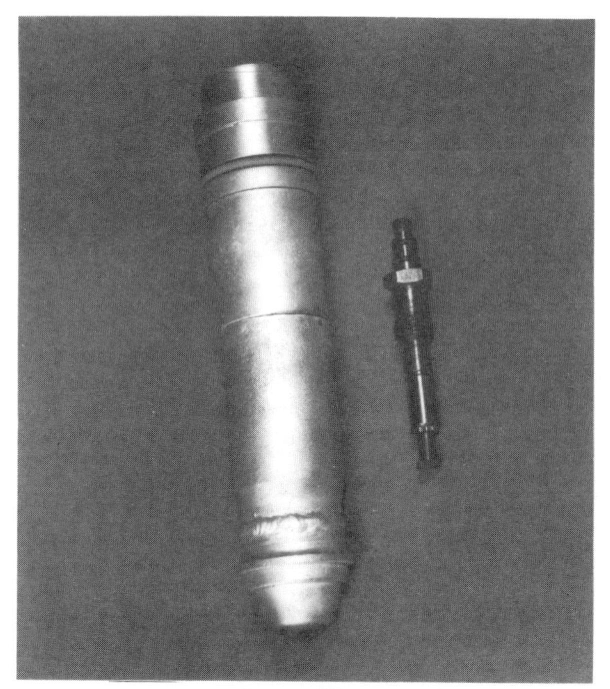

Figure 15
Pilot Oil Torch Cell and Injector

Figure 16
Installation of Pilot Oil Torch Cells
in LSB-6 Cylinder Heads

Load acceptance for the gas-diesel CleanBurn™ system is excellent and equals that of the conventional 6% pilot oil system. Variable ignition timing for the gas-diesel CleanBurn™ system is offered for non base-loaded installations to provide peak efficiency throughout the load range.

FIELD CONVERSION

To achieve "proof of concept" of the gas-diesel CleanBurn™ engine on a large field operating cogeneration unit, two existing LSVB-20GDT engines rated 6,000 KW, 200 BMEP were selected for the initial field conversions. Heads were removed, reconditioned and remachined for the torch cells; fuel and control system modifications assembled and installed; turbocharger upgraded and reinstalled; and the engine restarted within a three week time frame. A formal field test using a third party emissions measurement firm was performed the week after initial start-up of the first unit. Test results are gratifying, with NOx levels reduced to levels seen in the R&D engine, some reduction in CO levels and slight offset of hydrocarbons. The exhaust stacks are nearly totally clean with opacity improved from the 10-20% range of the non-CleanBurn™ gas-diesel engine, to 0 and 5% readings taken by a trained third party observer. Long-term reliability of the hardware appears to be excellent, with over 10,000 total hours logged.

CONCLUSION

The development of the gas-diesel CleanBurn™ engine is considered to be a major step forward in the advancement of emissions technology. It is expected that this is the new BACT level of emissions performance for dual-fuel engines.

ACKNOWLEDGMENTS

The authors would like to thank the management of Cooper-Bessemer Reciprocating Products Division, Cooper Industries, for permission and support to publish this paper. Recognition for contribution should go to the Research and Development Laboratory staff, Engineering Design personnel, and Jack Kimberley, fuel system consultant.

LSB-6 80°F AMBIENT
Gas-Diesel CleanBurn™ Mode 200 BMEP

Line	Detail	% PO	AMP	NOx	HCt	BSFC	FP	DEV	Smoke % Opacity	PTT
19	20° PC 200 BMEP	0.9%	67	1.36	3.9	6250	1266	48	< 5%	1021
20	20° PC 200 BMEP	0.9%	70	0.90	4.9	6330	1277	44	< 5%	991
21	23° PC 200 BMEP	0.9%	68	1.50	3.4	6210	1366	37	< 5%	942

Figure 17
Summary of Gas-Diesel CleanBurn™ Data

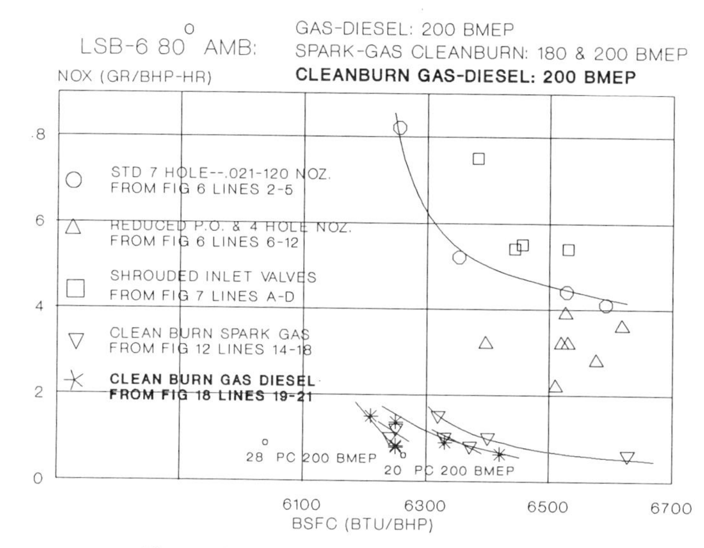

Figure 18 NOx/BSFC Plot for Gas-Diesel, Spark-Gas CleanBurn™, and Gas-Diesel CleanBurn™ Modes

METRIC EQUIVALENTS

Readers more familiar with metric units are asked to use the following information to convert the non-metric units used in this paper:

Nonmetric	Multiplied By	Yields Metric
Btu	1055.06	J
hp	745.70	W
in.	0.0254	m
psi	6.894.76	Pa or N/m^2
sq. in.	0.006452	m^2
(°F-32)	0.5556	°C
LB	0.4536	Kg

LSB-6 80°F AMBIENT
Gas-Diesel Mode - 200 BMEP

Line	Detail	% PO	AMP	NOx	HCt	BSFC	FP	DEV	Smoke % Opacity	PTT
1	Std. Gas-Diesel	5.7%	55	11.5	1.0	6231	1301	49	20	1207
21	CleanBurn™ Gas-Diesel	0.9%	70	0.90	4.9	6330	1277	44	< 5%	991

Figure 19
Comparison of Standard Gas-Diesel
to CleanBurn™Gas-Diesel

REFERENCES

1. Schaub, F.S., Hubbard, R.L., "A Procedure for Calculating Fuel Gas Blend Knock Rating for Large-Bore Gas Engines and Predicting Engine Operation." ASME Paper 85-DGP-5.

2. Schaub, F.S., Beightol, K.V., "NOx Reduction Methods for Large Bore Diesel and Natural Gas Engines." ASME Paper 71-WA/DGP-2.

3. Danyluk, P.R., Schaub, F.S., "Emission Reduction by Combustion Modification in Two Stroke Spark-Ignited Gas Engines and by Catalytic Conversion." ASME Paper 81-DGP-7.

4. Kaiser, R., "NOx Reduction Experience, Past and Planned, in Transamerica Delaval-Enterprise Engines." Presented at the American Gas Association Transmission Conference, Seattle, Washington, May 2, 1983.

5. Wilson, R.P. Jr., Mendillo, J.V., Gott, P.G., Danyluk, P.R., Schaub, F.S., Wasser, J.H., "Single-Cylinder Tests of NOx Control Methods for Spark-Gas Engines." ASME Paper 82-DGP-27.

6. Wilson, R.P. Jr., Mendillo, J.V., Genot, A., Bachelder, D.L., Wasser, J.H., "Single-Cylinder Tests of Emission Control Methods for Medium-Speed Diesel Engines." ASME Paper 82-DGP-28.

ICE-Vol. 15, Fuels, Controls, and Aftertreatment
For Low Emissions Engines
ASME 1991

OPTIMIZATION OF DIESEL FUEL INJECTION AND COMBUSTION SYSTEM PARAMETERS FOR LOW EMISSIONS USING TAGUCHI METHODS

T. P. Gardner
Ford Motor Company
Dearborn, Michigan

ABSTRACT

Taguchi design of experiment techniques have been successfully applied to optimize some key fuel injection and combustion system parameters for low emissions using a single cylinder, medium-duty, DI diesel engine. The relative effects of changes in the fuel spray cone angle, number of holes, nozzle hole area, nozzle protrusion, injection timing, compression ratio, and swirl level on exhaust emissions were quantified using statistical analysis, and the optimum parameter settings for low emissions were determined. The effect of EGR on the NOx/Particulate trade-off was also investigated. Results from the optimization experiments showed a significant reduction in both particulate and NOx emissions toward the 1994 Federal Standards.

INTRODUCTION

Most diesel manufacturers have managed to achieve the 1991 Heavy-Duty Diesel emission standards of 0.25 gm/bhp-hr particulates and 5.0 gm/bhp-hr NOx with relatively moderate improvements in combustion technology (fuel injection system optimization, combustion chamber modification, etc.) and in lube oil consumption control [2,3,4]. However, major improvements in in-cylinder combustion, lube oil consumption, fuel properties, and possibly exhaust aftertreatment devices may be required to further reduce particulates to 0.10 gm/bhp-hr at the same NOx level for the 1994 standards.

Many factors influence the formation of exhaust emissions in a direct-injection diesel engine. The fuel injection characteristics such as injection pressure, injection timing, injection rate, and nozzle geometry, play a significant role in the formation of particulates and NOx during the combustion process. The combustion chamber design including piston bowl shape, compression ratio, surface-to-volume ratio, and swirl level can have a significant impact on the formation of hydrocarbon and smoke emissions [5,6,7,8].

To determine the optimum combination of these various engine design parameters to achieve the lowest possible emissions using conventional "one-at-a-time " methods would require a large number of experiments which could be very time-consuming and costly. As an alternative, the Taguchi method [21] combines experimentation with statistical analysis to optimize several factors simultaneously and requires a much smaller number of experiments- Thus, reducing the time as well as the costs.

The purpose of this paper is to present the results from the application of the Taguchi method to optimize some key combustion and fuel injection system parameters for low emissions, and to show the potential benefits of using this method to reduce development time for 1994 diesel engines and beyond. The term "optimization" as used in this paper refers to the "relative" optimization of selected engine parameters over a specified limited range as opposed to the "absolute" optimization of the engine parameters.

THE TAGUCHI METHOD

The application of Taguchi design of experiments methods to investigate diesel combustion and emissions has been very limited [9,13]. Taguchi design of experiments methods are most extensively used to determine the parameter values or settings required to achieve the desired output function. Conventional design of experiments deals with averages only, while Taguchi design of experiments deals with averages and variability. Diesel engines must be designed and developed to meet emission values below the standards to allow for variability in product and process, and deterioration during useful life.

The Taguchi method can be used to determine the factors which affect the average emission levels as well as those which affect the variation. Additionally, the Taguchi method can be used to investigate the effects of interactions between the various factors which may be important. These important interactions can be easily missed when using conventional methods. For some of the investigations a combination of both conventional and Taguchi methods of testing were used depending on the kind of information that was required. The details of the Taguchi method can be found in [12,13].

Signal-to-Noise-Ratio. Taguchi introduced the concept of signal-to-noise in multivariate experimentation. The signal-to-noise (S/N) ratio is an objective measure of performance that takes both the mean and variation into account. The S/N ratio is used to evaluate the quality of an output characteristic which, in this experiment, was exhaust emissions. The method of calculating the signal-to-noise ratio differs depending on whether a larger response, a smaller response, or an "on-target" response is desirable. For emissions, the smaller-the-better method was selected and the S/N ratio was calculated using;

$$S/N = -10 \log\left(1/n \sum y_i^2 \right) \qquad [1]$$

where, n is the number of responses, and y_i is the response characteristic at level i. Note that the smaller the values of y_i (emission responses), the larger the S/N ratio. Therefore, the goal when using this method was to achieve the highest S/N ratio (i.e. lowest emissions) possible.

EXPERIMENTAL APPARATUS

Engine Description. A single cylinder, four-valve, overhead-cam, direct-injection diesel engine was used for this study. The engine had a 112 mm x 132 mm bore and stroke. A swirl-assisted quiescent combustion system with a centrally located fuel injector and piston bowl was used to enhance fuel-air mixing. The intake port design consisted of a high swirl port and a directed port with a built-in butterfly valve to vary air velocities during intake. The engine used a two-piece articulated piston (cast iron crown and aluminum skirt) with a reentrant bowl design yielding a nominal compression ratio of 17.5:1.

Fuel Injection System. Injection pressure, spray hole configuration, injection duration, and injection timing are those relevant fuel-air mixing parameters, which are determined solely by the design of the fuel injection system. An electro-mechanical, cam-driven unit injection system capable of injection pressures up to 1500 bar at full load was used in this study. Fuel nozzle opening pressure was set to approximately 280 bar. Injection timing and duration were electronically controlled by adjusting potentiometer knobs on the control panel. More than 60 fuel injector nozzles with different combinations of cone angle, number of holes, hole areas, and hole diameters were available for this experimental program.

Test Facility and Instrumentation. Engine speed and load (Torque) was set and maintained using a 300 HP solid state electric dynamometer. Air flow to the engine was supplied through a calibrated critical-flow measuring orifice with the capacity to allow turbo-charging up to 75 in Hg. Heaters were installed in the inlet plenum to control the inlet air temperature. Electro-mechanical actuators were used to control the swirl level generated in the intake ports. Fuel flow rate was measured using an AVL Type 703 Dynamic Fuel Consumption Measurement System with a micro-processor controlled Evaluating and Display unit.

For the exhaust gases, hydrocarbons were measured by Heated Flame Ionization Detector (HFID) analyzers, CO and CO_2 were measured using Non-Dispersive Infrared analyzers, and NOx emissions were measured using Chemiluminescence analyzers.

Particulate samples were obtained using a very efficient and compact mini-dilution tunnel design which uses only a fraction of the total engine exhaust to obtain accurate and repeatable particulate samples. The particulate samples were collected on 142 mm diameter fluorocarbon type filters having a nominal porosity of 1 micron. Three consecutive 10 minute particulate samples were taken for each test point. Exhaust smoke was measured using an AVL Type 409 Smoke Meter and Sampler Unit.

The data acquisition system (DAS) consisted of an IBM AT computer based system coupled with a Data Precision (DP) 6000 unit, and a Keithly 500 measurement and control system. The DAS system provided the capability to record both steady-state and transient engine data and allowed up to 100 engine input parameters to be monitored simultaneously.

DESCRIPTION OF THE EXPERIMENT

Prior to commencing the optimization experiments, a series of "screening" tests were conducted to determine the appropriate ranges of cone angle, injection timing, nozzle protrusion, and swirl level to use for this particular engine. The primary purpose in running these screening tests was to limit the range on the parameter settings to those which produced emission levels close to our program objectives. Once the appropriate ranges were determined, detailed experiments were conducted to optimize the parameters at different operating conditions and to assess the potential of this engine design to meet the 1994 emissions objectives.

Table 1
Parameter Listing

Parameters		Level 1	Level 2
A:	Cone Angle	Narrow	Wide
B:	No. of Holes	Few	More
C:	Nozzle Protrusion	Long	Short
D:	Swirl Level	Low	High
E:	Nozzle Area	Small	Large
F:	Compression Ratio	Low	High
G:	SOC Timing	Retard	Advance

Interactions		
A x B:	Cone Angle	x No. of Holes
A x E:	Cone Angle	x Nozzle Area
A x C:	Cone Angle	x Nozzle Protrusion
A x F:	Cone Angle	x Compression Ratio
A x G:	Cone Angle	x SOC Timing
A x D:	Cone Angle	x Swirl Level
B x C:	No. of Holes	x Nozzle Protrusion
D x G:	Swirl Level	x SOC Timing

Parameters Investigated. Seven parameters were selected for the optimization experiment as shown in Table 1. These parameters were considered to play an important role in the fuel-air mixing process and were expected to have a significant influence on the exhaust emissions;

A: Cone Angle. The wider the spray cone angle, the better the entrainment of air around the fuel plumes, thus, promoting better mixing.

B: Number of Holes. The number of fuel spray holes can significantly affect HC, NOx and smoke emissions depending on the combustion chamber design and swirl level.

C: Nozzle Protrusion. The distance that the nozzle tip protrudes into the combustion chamber affects the location where the fuel hits the piston bowl and walls which can also influence fuel-air mixing.

D: Swirl Level. Air swirl undoubtedly has the most directly observable influence on the fuel-air mixing and combustion process. Increased swirl is generally known to reduce particulates, hydrocarbons, and smoke, but often increases NOx.

E: Nozzle Area. The nozzle hole area affects the size of the fuel droplets produced and the degree of atomization. Smaller fuel droplets generally improves the mixing process as well as the combustion process.

F: Compression Ratio. The compression ratio (CR) plays an important role during the ignition delay period. Increasing CR results in higher compression temperatures, which tends to shorten ignition delay and increase NOx emissions.

G: SOC Timing. Start-of-Combustion (SOC) timing is known to have a considerable effect on NOx emissions. NOx emissions can be drastically reduced by retarding SOC timing. Of course, there is a fuel economy penalty associated with the retarded timing.

Two levels covering the upper and lower limits of the pre-determined operating range for this engine were selected for each parameter. In addition to these seven parameters, eight different interactions, also shown in Table 1, were selected for investigation.

Although the primary focus of this experiment was to reduce particulate and NOx emissions, hydrocarbon and smoke emissions were also measured to determine how these would be affected by the changes in the parameters. For all test points, low sulfur (.05% content) fuel was used to help reduce total engine-out particulates to levels required for 1994.

The L_{16} Design Array. To investigate the seven parameters and eight interactions listed in Table 1 requires a total of 15 degrees-of-freedom - 7 for main effects and 8 for interactions. Therefore, the L_{16} orthogonal design array was selected for this experiment. Figure 1 shows the linear graph for the L_{16} array. The numbers shown on this graph represent the 15 columns used for this array. The column numbers for the seven main parameters are shown at the nodal points and the column numbers for the eight interactions are shown on the lines connecting the nodal points.

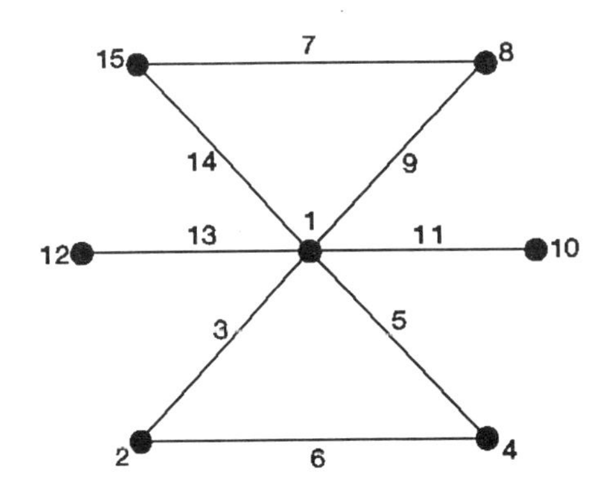

Figure 1 - Linear Graph of L_{16} Array

Table 2 shows the actual experimental layout of the L_{16} design array. This array represents a test matrix of 16 different engine configurations with various combinations of cone angles, number of holes, nozzle protrusions, swirl levels, hole areas, compression ratios, and injection timings.

Table 2
Experimental Layout of the L_{16} Design Array

No.	Cone Angle 1	No. Holes 2	Noz. Prot 4	Swirl 8	Noz. Area 10	CR 12	SOC Timing 15
1	Narrow	Few	Long	Low	Small	Low	Ret
2	Narrow	Few	Long	High	Large	High	Adv
3	Narrow	Few	Short	Low	Small	High	Adv
4	Narrow	Few	Short	High	Large	Low	Ret
5	Narrow	More	Long	Low	Large	Low	Adv
6	Narrow	More	Long	High	Small	High	Ret
7	Narrow	More	Short	Low	Large	High	Ret
8	Narrow	More	Short	High	Small	Low	Adv
9	Wide	Few	Long	Low	Small	Low	Adv
10	Wide	Few	Long	High	Large	High	Ret
11	Wide	Few	Short	Low	Small	High	Ret
12	Wide	Few	Short	High	Large	Low	Adv
13	Wide	More	Long	Low	Large	Low	Ret
14	Wide	More	Long	High	Small	High	Adv
15	Wide	More	Short	Low	Large	High	Adv
16	Wide	More	Short	High	Small	Low	Ret

Test Conditions. The speed/load test matrix of the EPA heavy-duty transient cycle [14] was approximated by 8 steady-state modes, as shown in Table 3. Cyclic power and emission results were calculated as the sum of the weighted modal values shown in this table.
To reduce the number of data points required for the optimization experiment, only three speed/load conditions, modes 4, 5, and 8, were run for each of the 16 engine configurations. Mode 4 represented a low speed, medium load point where carbon particulates were relatively high. Mode 8 represented a high speed, high load point where, again, carbon particulates were usually high. Also, mode 8 was selected because it represented nearly 1/3 (30.5%) of the total 8-mode cycle. Mode 5 represented a high speed, light load point where the lube oil contribution to particulates was usually high. Emission results for a total of 48 data points were recorded for this experiment.

RESULTS AND ANALYSIS

Table 4 shows the emission results obtained from the L_{16} experiment for modes 4, 5 and 8. Note that "indicated" rather than "brake" specific emission values are shown. Since one of the objectives of this study was to assess the potential of this engine design to meet the 1994 emission standards, a weighted average, based on the 8-mode weighing factors, was used for statistical analysis of the emission results. A computer program [15] was developed to perform the statistical calculations required for the analysis. The weighted emission averages were accounted for by inputting an appropriate number of repetitions into the computer program for each mode normalized to mode 5. For example, mode 5 represented 6.7 percent of the 8-mode cycle which corresponded to one repetition. Mode 4 represented 7.1 percent of the cycle, hence, one repetition was used for mode 4. Mode 8, however, represented 30.5 percent of the cycle, therefore, five repetitions were input for mode 8.

Table 3
8-Mode Steady State Test Approximations of the EPA Heavy-Duty Transient Cycle

Mode	% Full Speed	% Full Load	Weighing Factor
1	Low Idle	1.0	.001
2	10.0	25.0	.021
3	22.0	63.0	.036
* 4	32.0	86.0	.071
* 5	99.0	18.0	.067
6	95.0	40.0	.186
7	95.0	70.0	.313
* 8	90.0	95.0	.305

Table 4
Emission Results from L16 Experiment (Modes 4, 5, and 8)

	ISPART (gm/ihp-hr)			ISNOX (gm/ihp-hr)			ISHC (gm/ihp-hr)			SMOKE (bosch)		
No	4	5	8	4	5	8	4	5	8	4	5	8
1	.033	.075	.083	4.35	2.85	3.19	.106	.221	.056	0.6	0.3	0.9
2	.047	.102	.087	9.38	5.88	5.51	.081	.153	.052	0.4	0.4	0.6
3	.052	.057	.070	6.73	4.44	4.08	.109	.214	.072	0.5	0.2	0.6
4	.054	.123	.071	5.45	3.29	3.76	.100	.214	.084	0.5	0.2	0.7
5	.042	.079	.109	6.67	5.59	4.73	.077	.139	.049	0.2	0.05	1.2
6	.038	.095	.057	6.84	3.69	4.82	.101	.147	.054	0.4	0.1	0.1
7	.048	.053	.079	4.66	3.79	4.51	.100	.164	.038	0.6	0.1	0.8
8	.042	.050	.047	8.89	5.45	5.79	.091	.163	.070	0.0	0.5	0.3
9	.093	.111	.084	8.15	4.03	4.06	.096	.189	.074	0.7	0.2	0.6
10	.066	.100	.108	5.15	3.43	3.87	.074	.216	.066	0.6	0.3	0.7
11	.116	.126	.118	4.46	3.00	3.44	.096	.221	.073	1.3	0.2	1.1
12	.049	.077	.041	9.09	3.32	5.11	.077	.113	.053	0.5	0.1	0.4
13	.060	.114	.083	4.70	3.72	3.95	.084	.171	.092	0.4	0.1	0.7
14	.036	.171	.064	8.84	6.43	5.19	.085	.137	.053	0.3	0.5	0.4
15	.070	.050	.066	7.54	6.02	5.75	.096	.178	.072	0.7	0.1	0.5
16	.046	.150	.044	5.52	3.47	4.01	.092	.153	.046	0.3	0.05	0.3

Table 5
Analysis of Variance (ANOVA)
From L_{16} Experiment

	% Contribution			
Parameter	PART.	NOx	HC	SMOKE
Cone Angle	5.3	----	----	----
No. Holes	5.9	11.3	15.2	15.5
Protrusion	19.6	----	----	----
Swirl Level	10.7	11.6	11.7	41.7*
Nozzle Area	----	----	----	11.8
Comp Ratio	4.1	1.9	----	----
SOC Timing	8.1	65.4*	11.7	----
Angle x Area	26.3*	----	2.2	9.8
Holes x Prot	----	----	----	----
Angle x Prot	----	----	15.1	----
Angle x CR	6.7	----	8.4	10.5
Angle x SOC	6.9	----	----	----
Angle x Hole	----	----	5.8	----
Swirl x SOC	----	----	3.9	----
Angle x Swirl	----	----	16.4*	----
Pooled Error	6.4	9.8	9.6	10.7

ANOVA Results. Table 5 shows the results from an analysis of variables (ANOVA) for the four emission responses. These numbers indicate the relative percent contribution of each parameter to the total variation observed in the emission response when changing from one level to another. For example, 26.3 percent of the total variation in particulates was due to the interaction between cone angle and nozzle area when changing from level 1 to level 2. This interaction apparently had the most significant effect on particulate. Nozzle protrusion was second in significance contributing 19.6 percent, followed by swirl level at 10.7 percent. Parameters contributing less than 2 percent to the total variation were "pooled" together in the error term.

For NOx, SOC timing was the dominate parameter contributing 65.4 percent to the total variation. Swirl level and the number of fuel spray holes also had significant effects on NOx emissions contributing 11.6 and 11.3 percent, respectively.

Hydrocarbon emissions were mainly influenced by the interactions between cone angle and swirl, and between cone angle and nozzle protrusion. The number of spray holes, and SOC timing also influenced HC emissions.

Smoke emissions were mainly controlled by swirl level (41.7 %) followed by the number of spray holes (15.5 %) and nozzle area (11.8 %).

Table 6
S/N Ratios for each
Parameter Level

	PART		N O x		H C		SMOKE	
Parameters	1	2	1	2	1	2	1	2
A: Cone Angle	22.90	21.89	-13.61	-13.46	20.81	20.85	5.76	5.70
B: No. of Holes	21.87	22.93	-12.94	-14.13	20.22	21.44	4.10	7.37
C: Nozzle Protr.	21.45	23.35	-13.53	-13.54	20.97	20.69	5.34	6.12
D: Swirl Level	21.69	23.10	-12.93	-14.14	20.30	21.37	3.07	8.40
E: Nozzle Area	22.58	22.22	-13.31	-13.76	20.63	21.04	7.17	4.29
F: Compress Ratio	22.85	21.95	-13.26	-13.81	20.84	20.83	5.76	5.71
G: SOC Timing	21.78	23.02	-12.14	-14.94	20.30	21.37	5.34	6.13
Interactions								
A x B:	22.39	22.41	-13.44	-13.63	20.45	21.22	5.71	5.75
A x E:	23.49	21.31	-13.58	-13.49	20.57	21.09	7.05	4.42
A x C:	22.46	22.34	-13.61	-13.46	21.44	20.23	5.86	5.61
A x F:	21.83	22.96	-13.39	-13.69	20.37	21.29	4.37	7.09
A x G:	22.97	21.83	-13.69	-13.39	21.00	20.67	6.35	5.11
A x D:	22.43	22.37	-13.27	-13.81	21.47	20.20	5.12	6.35
B x C:	22.61	22.19	-13.65	-13.42	20.81	20.86	5.95	5.51
D x G:	22.58	22.22	-13.63	-13.44	21.16	20.51	5.22	6.24

In addition to the ANOVA table, the S/N ratios at each parameter level was calculated and are listed in Table 6. Recall that, the parameter levels which gave the highest S/N ratio corresponded to the level which produced the lowest emissions. Hence, one can select the best (optimum) parameter settings for each emission response from Table 6 by simply selecting the level with the highest S/N ratio.

Optimum Parameter Selection. To obtain a clear idea of the experimental results, the effect (in terms of S/N ratio) of the four most significant parameters for each emission response is graphed in Figure 2. The parameters are arranged so that the most significant is on the left. Since the higher S/N ratio was more desirable, one could easily select the parameter level which gave the best results from this graph. Table 7 summarizes the optimum parameter settings selected for each emission response. Note that the term "optimum" reflects only the optimum levels of the parameters as defined by this experiment.

Table 7 clearly shows the conflicting parameter settings which are required to achieve the lowest emissions. To achieve low particulates one would use more fuel spray holes, high swirl, and advanced SOC timing. For low NOx, one would use fewer fuel spray hole, low swirl, and retarded SOC timing. Low HC emissions would require a wide cone angle with a long nozzle protrusion while low smoke would require a narrow cone angle with a short protrusion.

To satisfy the conflicting requirements for all four emissions, a trade-off setting was selected. A narrow cone angle, small

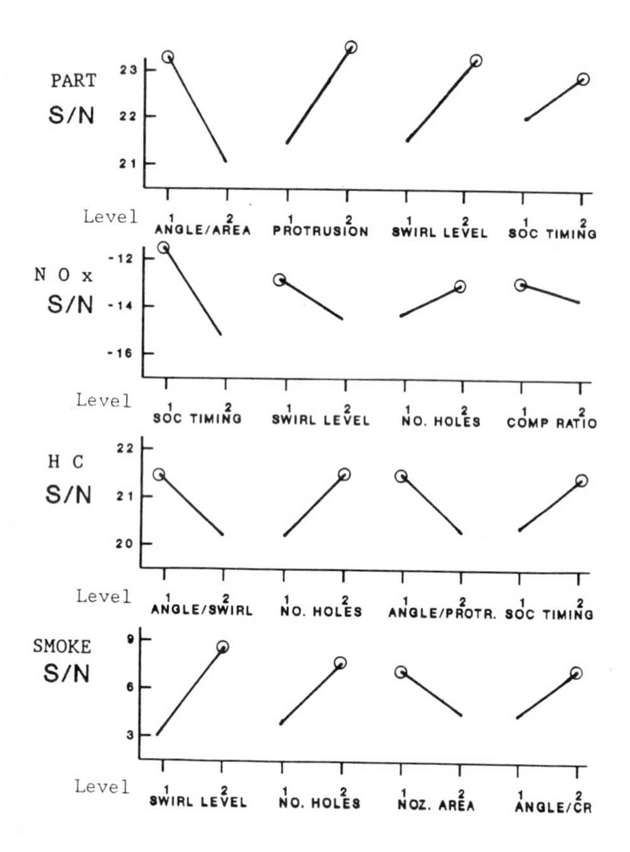

Figure 2 - Graph of Significant Effects

Table 7
Optimum Parameter Settings
Based on Highest S/N Ratios

Parameter	PART.	NOx	HC	SMOKE	Trade-Off
Cone Angle	Narrow	----	Wide	Narrow	Narrow
No. Holes	More	Few	More	More	More
Protrusion	Short	----	Long	Short	Short
Swirl	High	Low	High	High	High
Noz. Area	Small	----	----	Small	Small
Comp Ratio	Low	Low	----	----	Low
SOC Timing	Adv	Ret	Adv	Adv	Ret

nozzle area, and a short protrusion was selected to satisfy the particulate and smoke requirement. The nozzle with more holes was selected to satisfy the HC requirement. High swirl was selected to satisfy the smoke and HC requirement, and retarded SOC timing with a low compression ratio was selected for the NOx requirement.

Prediction of Emissions at Optimum Conditions. Since the optimum (trade-off) condition shown in Table 7 was not tested in the L_{16} experiment, one would like to know what improvement in the emission results could be expected at this condition? To estimate the emission responses at the trade-off conditions using the experimental results, the following simple equation was used (13);

$$OPT = \overline{T} + \sum(\overline{X} - \overline{T}) \pm \sqrt{F\ Ve\ \frac{1}{Ne}} \qquad [2]$$

where, $\overline{T}$ = overall mean of the experiment
$\overline{X}$ = mean of a significant effect
Ve = variance due to error
Ne = effective d.o.f.
= total d.o.f. / (1 + d.o.f. for signif. effects)
F = F statistic for v1 and v2
v1 = sum of d.o.f. for signif. effects
v2 = total d.o.f. for the experiment

This equation assumes that each factor is independent and that no significant interactions exists. To allow for the possibility of an over-estimate due to error of variances, only parameters which had a strong effect on the emission response were used in calculating the estimate.

The optimum (trade-off) parameter settings from Table 7 can be designated as;

A1 B2 C2 D2 E1 F1 G1

where, A1 represents cone angle level 1, B2 represents number of holes level 2, C2 represents nozzle area level 2, etc. From the ANOVA Table 5, the strong effects for each emission response were determined to be;

ISPART : A1xE1 C2 D2 G1

ISNOx : G1 D2 B2

ISHC : A1xD2 B2 A1xC2 D2 G1

Smoke : D2 B2 E1 A1xF1 A1xE1

Using equation [2], and Table 6, estimates of the emission responses at the optimum settings were calculated and are shown in Table 8. Details of the calculations can be found in reference [9].

Table 8
Comparison of Predicted and Actual
S/N Ratios Using the Optimum Settings

Emissions	Predicted (95% Confidence)	Actual Results
PART	23.6 to 25.4	24.6
NOx	-12.4 to -14.3	-13.0
HC	19.4 to 21.0	20.8
Smoke	9.6 to 13.3	13.3

Confirmation Experiments. To test the accuracy of the above predictions, the engine was configured using the optimum (trade-off) parameter settings, and two emission samples were taken at modes 4, 5, and 8 test conditions. The weighted S/N ratios were calculated for each emission response and compared with the predicted values calculated using equation (2). Table 8 shows that the actual S/N ratios fell well within their predicted ranges which indicates good reproducibility in the results. Had these not compared within the 95% confidence range, the experimental results would have been suspect-Either a significant factor or interaction may have been missed, or the parameter levels may have been set too narrowly to detect the effect of changes in the levels.

8-Mode Steady-State Results. To assess the potential of this engine design to achieve the 1994 emission objectives, three engine configurations were tested using the 8-mode steady-state simulation. The results from these test are shown in Table 9. Configuration #1 represented the optimum (trade-off) parameter settings as determined from the statistical analysis. The cycle averaged brake specific particulate (BSPART) and brake specific NOx (BSNOx) emissions calculated for this case were .078 gm/bhp-hr and 5.36 gm/bhp-hr, respectively. This configuration satisfied the particulate target of 0.08 gm/hr-hr, but exceeded the NOx target of 4.5 gm/bhp-hr.

Configuration #2 was similar to #1 except that the swirl level was reduced slightly from high to medium. This reduction in swirl level resulted in a decrease in BSNOx to 4.88 gm/bhp-hr, but also, an increase in BSPART to 0.127 gm/bhp-hr.

Table 9
Results from Steady-State 8-Mode Simulation

Parameter	Run #1	Run #2	Run #3
Cone Angle	Narrow	Narrow	Narrow
No. Holes	More	More	Few
Nozzle Area	Small	Small	Small
Protrusion	Short	Short	Short
Comp Ratio	Low	Low	Low
Swirl Level	High	Med	High
SOC Timing	Retard	Retard	Variable
BSPART (.08)	.078	.127	.078
BSNOx (4.5)	5.36	4.88	4.60

For configuration #3, the swirl level was reset to high, the number of nozzle spray holes was reduced, and a variable SOC timing was used to control NOx. This configuration resulted in a BSPART of 0.78 gm/bhp-hr and BSNOx of 4.60 gm/bhp-hr which was very close to the 1994 emission objectives. Note, however, these results were obtained under steady-state conditions and that the EPA certification procedure is a transient procedure.

To summarize, the results showed that the single cylinder engine design used in this study has the potential to achieve the 1994 emission objectives without the use of an aftertreatment device. Of course, this is highly dependent on a further reduction in NOx emissions as well as particulates to allow for variability in production.

Effect of EGR on NOx and Particulates. The results from the previous section showed that a further reduction in NOx emissions will be needed in order to meet the 1994 objectives (4.5 gm/bhp-hr) at a particulate level of 0.08 gm/bhp-hr. Exhaust gas recirculation (EGR) is known to be an effective technique for reducing the maximum temperature reached during combustion by diluting the incoming fresh air charge with exhaust gases--thus reducing the formation of NOx. Therefore, a test was conducted to evaluate the benefits and/or penalties associated with using EGR to further reduce NOx emissions at selected speed/load conditions.

For this test, modes 4, 6, 7 and 8 conditions were selected from the 8-Mode steady-state test matrix. These modes were selected because they represented relatively high loading conditions where EGR would be most beneficial, and because they were the most heavily weighted modes in the 8-mode cycle (87.5%).

The test procedure for conducting this experiment consisted of setting up the base data point without EGR and measuring the CO2 concentration in exhaust (CO2exh). Then, the EGR valve was opened allowing the exhaust gases to mix with the inlet air. The concentration of CO2 in the intake port (CO2int) was monitored while adjusting the engine exhaust backpressure to maintain the desired EGR rate. The EGR flow rate was determined by;

$$\%EGR = \frac{CO2int - CO2backgr}{CO2exh - CO2int} \times 100 \quad [3]$$

where, CO2backgr was the CO2 concentration in the cell ambient conditions. The comparison of the emission results for the base engine with %5 EGR and without EGR are shown in Table 10.

Table 10 shows that using 5% EGR at Modes 4, 6, 7, & 8 resulted in a 23% reduction in NOx for the cycle but also increased particulates by 31%. These results agree with those obtained from similar EGR studies [2]. Smoke emissions also increased when using EGR.

Table 10
Comparison of Base Emissions With and Without EGR

Mode	BSPART (gm/bhp-hr)		BSNOx (gm/bhp-hr)		SMOKE (bosch)	
	BASE	5% EGR	BASE	5% EGR	BASE	5% EGR
4	.068	.076	6.33	3.83	.4	.6
6	.108	.158	3.80	3.14	.6	.8
7	.077	.103	4.12	2.83	.6	.7
8	.050	.077	4.54	3.42	.5	.8
Cycle	.080	.105	4.62	3.57		

Therefore, since the use of EGR to control NOx to acceptable levels would adversely affect particulates and smoke, other NOx control techniques (including aftertreatment) may be needed to achieved the 1994 objectives for the 7.8L engine.

CONCLUSIONS

Based on the results presented in this paper, the major conclusions can be summarized as follows;

1) The potential to achieve the 1994 heavy-duty diesel standards for NOx and Particulate emissions was demonstrated using a single cylinder DI diesel engine under steady-state conditions with low (.05%) sulfur fuel and without EGR or any aftertreatment devices.

2) The optimum (trade-off) settings for seven important combustion and fuel injection system parameters were determined using Taguchi design of experiments which resulted in particulate and NOx emission levels of 0.078 gm/bhp-hr and 4.6 gm/bhp-hr, respectively. However, no assessment was made as to what extent these results could be reproduced in a multi-cylinder engine under volume production conditions and through extended mileage.

3) The most significant parameters influencing particulates were found to be nozzle protrusion, swirl level, and the interaction between cone angle and nozzle area. NOx emissions were mainly governed by SOC timing, swirl level, and the number of spray holes. HC emissions were sensitive to the interactions between cone angle, protrusion, and swirl as well as the number of spray holes. Smoke emissions were mainly controlled by swirl level, number of spray holes, and the nozzle area.

4) Using 5 %EGR at selected 8-mode conditions resulted in a 23% decrease in the cycle averaged NOx and a 31% increase in particulates. Hence, other NOx control techniques (including aftertreatment) may be needed to maintain the emissions below the 1994 standards over the useful life of the engine.

ACKNOWLEDGEMENTS

The author wishes to express his sincere appreciation to Messrs. Fred Bremfoerder and David Vollmer of Ford NewHolland for their support and cooperation during this project, and to Messrs. Phillip Dingle and Kerry Transit of Lucas CAV for their assistance with the fuel injection system. The helpful advice and assistance of Mr. Charles Hunter, Research Staff, was greatly appreciated. Special thanks to Mr. Mark Seaman, Technician, for running the experiments.

REFERENCES

1. "Federal Heavy Duty 5.0 gm/bhp-hr NOx and 0.25 gm/bhp-hr Particulate Standard," 50 Federal Register 10653, 40CFR, 86.091-11, March 15, 1985.
2. P. Zelenka, W. Kriegler, P.L. Herzop, and W.P. Cartellieri, "Ways Toward the Clean Heavy-Duty Diesel", SAE Paper No. 900602, 1990.
3. A.P. Gill, "Design Choices For 1990's Low Emission Diesel Engines", SAE Paper NO. 880350, 1988.
4. W.P. Cartellieri and P.L. Herzog, "Swirl Supported or Quiescent Combustion for 1990's Heavy-Duty DI Diesel Engines - An Analysis", SAE Paper No. 880342, 1988.
5. C.A. Amann and D.C. Siegla, "Diesel Particulate--What They Are and Why", GM Research Publication No. GMR-3672, 1981.
6. R.J. Hames, D.F. Merrion, and H.S. Ford, "Some Effects of Fuel Injection System Parameters on Diesel Exhaust Emissions," SAE Paper No. 710671, 1971.
7. J.H. Van Gerpen, C. Huang, and G.L. Borman, "The Effects of Swirl and Injection Parameters on Diesel Combustion and Heat Transfer," SAE Paper No. 850265, 1985.
8. T.J. Williams and M. J. Tindal, "Gas Flow Studies in Direct Injection Diesel Engines with Re-Entrant Combustion Chambers", SAE Paper NO. 800027, 1980.
9. C.E. Hunter, T.P. Gardner, and C.E. Zakrajsek, "Simultaneous Optimization of Diesel Engine Parameters for Low Emissions Using Taguchi Methods", SAE Paper No. 902075, 1990.
10. K.J. Springer, "Low Emission Diesel Fuel for 1991-1994", Advances in Engine Emission Controls Technology ICE - Vol.5, 1989.
11. G. Stumpp, W. Polach, N. Muller, and J. Warga, "Fuel Injection Equipment for Heavy Duty Diesel Engines for U.S. 1991/1994 Emission Limits", SAE Paper 890851, 1989.
12. Genichi Taguchi, "Introduction to Quality Engineering,", Kraus Inter. Publications, Whiter Plains, New York, 1986.
13. American Supplier Institute, Inc., "Introduction to Quality Engineering," Copyright 1987.
14. USA Code of Federal Regulations, "Protection of Environment Title 40, Part 86, Revised July 1, 1985 and Dec. 16, 1987.
15. Design of Experiments (DOE) Computer Program, developed by Quality Software Products, Inc., Copyright 1987.
16. T. B. Barker, "Quality by Experimental Design", Marcel Dekker, Inc., ASQC Quality Press, New York, Copyright 1985.

ICE-Vol. 15, Fuels, Controls, and Aftertreatment
For Low Emissions Engines
ASME 1991

THE EFFECT OF INJECTION TIMING, ENHANCED AFTERCOOLING, AND LOW SULFUR-LOW AROMATIC DIESEL FUEL ON LOCOMOTIVE EXHAUST EMISSIONS

Vernon O. Markworth
Department of Engine Research
Steven G. Fritz
Department of Emissions Research
Southwest Research Institute
San Antonio, Texas

G. Richard Cataldi
Research and Test Department
Association of American Railroads
Washington, D.C.

ABSTRACT

An experimental study was performed to demonstrate the fuel economy and exhaust emissions implications of retarding fuel injection timing, enhancing charge air aftercooling, and using low-sulfur, low-aromatic diesel fuel for locomotive engines. Steady-state gaseous and particulate emissions data are presented from two 12-cylinder diesel locomotive engines. The two laboratory engines, an EMD 645E3B and a GE 7FDL, are each rated at 1,860 kW (2,500 hp) and represent the majority of the locomotive fleet in North America. Each engine was tested for total hydrocarbons (HC), carbon monoxide (CO), oxides of nitrogen (NO_X), and particulate. Emissions were measured at three steady-state operating conditions; rated speed and load, idle, and an intermediate speed and load. Test results on the EMD engine indicate that a 4° injection timing retard, along with a low-sulfur low-aromatic fuel and enhanced aftercooling was effective in reducing NO_X from 10.5 g/hp-hr to 7.2 g/hp-hr; however, particulates increased from 0.15 g/hp-hr to 0.19 g/hp-hr, and fuel efficiency was 4.3 percent worse. Similar observations were made with the GE engine. This paper gives details on the test engines, the measurement procedures, and the emissions results.

NOMENCLATURE

AAR Association of American Railroads
ASTM American Society for Testing and Materials
bhp brake horsepower
°C degrees Celsius
CARB California Air Resources Board
CO Carbon Monoxide
CO_2 Carbon Dioxide
dc direct current
EMD Electro-Motive Division of General Motors Corporation
EPA United States Environmental Protection Agency
°F Degrees Fahrenheit
g grams
GE Transportation Systems Division of General Electric Company
HC Hydrocarbons (total)
hr hour
kW kilowatt
NO_X Oxides of Nitrogen
PM Particulate Matter
SAE Society of Automotive Engineers
SwRI Southwest Research Institute

INTRODUCTION

The program on which this paper is based is part of a multi-year locomotive engine research project funded by the Association of American Railroads (AAR) to characterize emissions from locomotive engines and to determine the effect of various control technologies on unburned hydrocarbons (HC), carbon monoxide (CO), oxides of nitrogen (NO_X), and particulates (PM). A previous study by Fritz and Cataldi [1] presented results of baseline locomotive engine emissions along with test results from a low sulfur fuel currently used in southern California.

The current regulatory climate regarding engine emissions suggests that locomotive engine emissions (both gaseous and particulate) will soon be regulated beyond existing visible smoke standards. Regulation will likely begin in California where continuing non-attainment of National Ambient Air Quality Standards in many areas has resulted in consideration of extending mobile source emission regulations to off-highway sources (including locomotives). Assembly Bill AB234 was enacted in 1987 by the California Legislature and required that the California Air Resources Board (CARB) to perform a study of locomotive emissions in the six non-attainment basins within California before any regulatory action could be taken. A contractor was selected to perform this study, and the final report [2] has recently been submitted to the CARB for final approval. This will clear the way for the rulemaking process for regulating locomotive emissions. The California Clean Air Act of 1988 called for a CARB staff report to develop recommendations for regulating off-highway emissions by the end of 1991. These regulatory recommendations are to include locomotives.

One recommendation to reduce locomotive exhaust emissions was presented in the CARB Locomotive Emissions Study [2]. It combined implementation of a 4° injection timing retard with the use of a low-sulfur, high cetane diesel fuel for yard and switcher locomotives. The experimental efforts described herein were performed in an attempt to quantify the emissions and fuel economy implications associated with these recommended modifications.

The Federal Clean Air Act Amendments of 1990 required that the Environmental Protection Agency (EPA) study the air quality impact of locomotive emissions in non-attainment areas. It did not, however, set specific engine emission levels. It did include federal pre-emption for new locomotives and locomotive engines which excluded California or any other state from regulating new locomotives. This left the door open for CARB to pursue regulations on existing locomotives. As part of EPA's efforts, an initial survey of the contribution of locomotive emissions to air pollution in non-attainment areas is to be performed within one year. If the study concludes that locomotive emissions are significant, regulation must be in place within five years. Initial efforts in this area have focused on the EPA revising their methodology in state implementation plan guidelines for quantifying locomotive engine emissions.

In Canada, the Canadian Council of Ministers of the Environment, in their Phase I Management Plan for Nitrogen Oxides and Volatile Organic Compounds [3] will require railroads to reduce their total NO_X by 14 percent on a nationwide basis from the base year of 1985 by the year 2005. Current (1985) NO_X emissions from all rail transport were estimated at 7 percent of the national total. The NO_X/VOC Management Plan - Phase I represents the first phase of a three-phase NO_X and VOC control program aimed at fully resolving ground-level ozone problems in Canada by the year 2005. The Phase I plan is not expected to be sufficient to meet Canada's air quality goals, so the railroad industry's emission targets for 2005 could be lowered in a future Phase. Because of the relatively small Canadian locomotive market, the government recognizes that they cannot exceed the stringency of U.S. regulations. After EPA and CARB act on new and existing locomotive engine regulations, the Canadian railroads could face more stringent regulations.

BACKGROUND

There are effectively only two basic engines represented in the U.S. locomotive population. One of these is the two-stroke cycle engine manufactured by Electro-Motive Division (EMD) of General Motors Corporation. The majority of Class I U.S. locomotives (approximately 70 percent) are powered by a variant of this engine. Nearly all of the remaining U.S. locomotives use the four-stroke cycle engine manufactured by the Transportation Systems Division of the General Electric Company (GE). Two engine designs manufactured by Caterpillar Inc. are being field tested in a limited number of locomotives. A variety of other engine designs can be found on older switcher locomotives owned by short-line railroads and private industrial railroads.

A very limited amount of locomotive emissions information exists in the public domain. The Southern Pacific Transportation Company (SP) performed a study to develop certification procedures for visible smoke from locomotives in 1973 [4]. A field study was performed on GE locomotives by SwRI in conjunction with GE and the SP in 1975 [5]. Several studies of locomotive engine emissions were performed by SwRI for the U.S. Department of Transportation, the Federal Railroad Administration and the EPA in the mid-1970's [6-10]. These studies focused on gaseous emissions and visible smoke (or percent opacity using light extinction smoke meters) and were performed on engines that are now two generations old. Most of these locomotives have been retired, upgraded, placed in switcher service, or sold to short line railroads where they are infrequently used.

A field study of in-service locomotive exhaust emissions was performed by the AAR in 1981 through 1983 [11]. Other studies by the AAR in the early and mid-1980s were performed primarily to determine engine performance while operating on alternative fuels, emulsions, or broadened specification diesel fuels. During some of these tests, gaseous emissions were recorded [12-16]. Since that time, however, significant engine upgrades in both the EMD and GE engine families have occurred, and a corresponding update of emission data is needed.

EXPERIMENTAL TEST PLAN

Description of Test Engines and Facilities

The EMD 12-645E3B engine shown in Figure 1 is a two-stroke cycle engine, while the GE 12-7FDL operates on the four-stroke cycle. In the SwRI locomotive engine test facility, the EMD is loaded by a dc generator and the GE by a dc rectified alternator. The dc power is absorbed by two sets of locomotive load grids equipped with cooling fans for heat dissipation. Both engines have their own cooling systems which are modified versions of the standard locomotive system. Pertinent specifications for each of the test engines are shown in Table 1.

Both engines are operated at conditions that simulate the notch operation of their respective line-haul locomotives. Locomotive engines operate only at specified speed/power combinations defined by throttle positions or "notches." Since the power output and speed are constant at each notch position, the fuel consumption rate is also constant and can be defined for each position. EMD and GE engines in line-haul service operate at eight power producing notches. There are also idle and, in some locomotives, low idle operating positions. In dynamic brake throttle notch position, traction motors are used to brake the train. The dc power generated during the dynamic brake operation is dissipated through resistive load grids located on the roof of the locomotive.

To facilitate laboratory testing, the notch positions were redefined in terms of speed and fuel mass consumption rate combinations. Notch 8 of each schedule represents rated speed and load of the engine. All other positions represent part-load conditions. To simplify emissions test work, only three notches were used. These are given in Table 2.

12-CYLINDER EMD-645E3

12-CYLINDER GE-7FDL

Figure 1. Locomotive Test Engines at SwRI

Table 1. Locomotive Test Engine Specifications

	EMD 12-645E3B	GE 12-7FDL
Number of Cylinders	12	12
Displacement L (in^3) / cyl	10.6 (645)	10.9 (668)
Bore and Stroke mm (in)	230 x 254 (9-1/16 x 10)	229 x 267 (9 x 10½)
Rated Speed (rpm)	904	1,050
Rated Power @ flywheel kW (bhp)	1,860 (2,500)	1,860 (2,500)
Compression Ratio	14.5:1	12.7:1
Cycle	2	4
Injection System	Unit Injector	Jerk Pump/ Nozzle

Table 2. Throttle Notch Schedules

	EMD 12-645E3B		GE 12-7FDL	
Notch Position	Engine rpm	Fuel Rate (kg/hr)	Engine rpm	Fuel Rate (kg/hr)
Idle	300	**	450	**
5	650	183	879	206
8	904	396	1,050	391
** - Observed at idle speed				

Gaseous Emission Measurement

Gaseous emission measurements during each steady-state test condition were obtained by sampling raw exhaust. Exhaust gases were analyzed for unburned hydrocarbons (HC), carbon monoxide (CO), oxides of nitrogen (NO_x), carbon dioxide (CO_2), and oxygen (O_2). Hydrocarbons were measured by a heated flame ionization detector unit built to specification as per SAE Recommended Practice J215. Carbon monoxide and carbon dioxide were measured by non-dispersive infrared analyzers in a system that conforms to SAE Recommended Practice J177a. Oxides of nitrogen were measured using a chemiluminscent analyzer. For HC, CO, CO_2, and NO_x, all measurement instrumentation conformed to configurations typically used for EPA regulatory purposes.

Particulate Emission Measurement

The generally accepted test procedure for measurement of diesel particulate is specified by EPA for on-highway diesel engines. This procedure requires dilution of the total exhaust flow. Because of the size of locomotive engines, this "full-flow" dilution approach is not practical. Consequently, particulate measurements were made with a "split-then dilute" system. This system samples a portion of the total exhaust flow and mixes it with filtered ambient air. Mixing occurs in the dilution tunnel prior to sampling the mixture for particulate. The dilution tunnel used in this work was 20 centimeters (8 inches) in diameter and approximately 5 meters (15 feet) long. Samples of particulate were collected on dual (series) 90 mm Pallflex T60A20 fluorocarbon-coated glass fiber filters. The temperature of the diluted exhaust stream at the entrance to the particulate sample probe was regulated to 49°C±3 by varying the amount of raw exhaust entering the tunnel. All particulate filters were conditioned and weighed in a temperature and humidity controlled chamber both before and after use.

Test Fuels Description

Two test fuels were evaluated in each of the 12-cylinder engines. For baseline measurements, an ASTM 2D diesel fuel was used. To assess the impact of low-aromatic, low-sulfur fuel on locomotive engine emissions, a second fuel was also tested. Both fuels were analyzed and the properties are given in Table 3. The low-aromatic, low-sulfur fuel was notably less dense, and has a substantially higher cetane number than the baseline fuel.

Table 3. Diesel Fuel Properties

Determinations		Test Method	Baseline	Low Aromatic - Low Sulfur
API Gravity @ 60°F		D1298	32.0	41.9
Flash Point °C (°F)		D93	64 (148)	77 (170)
Cloud Point °C (°F)		D2500	-8 (18)	-16 (4)
Viscosity @ 40°C (cSt)		D445	3.33	2.74
Sulfur (Wt%)		D2622	0.290	0.043
Cetane Number		D613	44.2	54.5
Heat of Combustion		D240		
Gross	MJ/kg		44.0	46.3
	BTU/lb		19,400	19,900
	BTU/gal		139,600	135,300
Net	MJ/kg		42.3	43.3
	BTU/lb		18,200	18,600
	BTU/gal		131,200	126,600
Carbon-Hydrogen Ratio		D3178		
% Carbon			87.3	85.7
% Hydrogen			12.2	14.2
Hydrocarbon Type		D1319		
Aromatics (%)			30.8	15.6
Olefins (%)			5.1	4.0
Saturates (%)			64.1	80.4
Specific Gravity			0.865	0.816
Distillation		D86		
		% Recovered	Temp. °C (°F)	Temp. °C (°F)
		IBP	188 (370)	191 (376)
		5	221 (430)	208 (406)
		10	237 (458)	216 (420)
		20	246 (474)	228 (443)
		30	254 (490)	242 (467)
		40	263 (506)	253 (488)
		50	280 (536)	263 (506)
		60	289 (552)	273 (524)
		70	299 (570)	283 (541)
		80	311 (592)	294 (562)
		90	329 (624)	311 (592)
		95	340 (644)	326 (619)
		EP	353 (668)	342 (648)

Duty Cycles

The engine test procedure used for the emissions measurements was essentially the same as has been developed for performance measurements. The engines were stabilized at rated conditions (Notch 8) for 2 hours prior to testing. One hour was spent at each notch position, with the first half hour being stabilization time. The emissions measurements took up to a full half hour at each test point.

The engine test procedure used for the emission measurements originally used a full-mode cycle and calculated the composite emissions using weighting factors for a line haul cycle provided by GE [1]. The full-mode test included the eight discrete power producing throttle positions or "notches", idle, low idle, and two dynamic brake notch positions. This cycle resulted from many studies performed by GE and EMD to determine actual operating modes of road locomotives.

Each emission test using the "full-mode" cycle required 10 to 12 hours to complete, and since multiple runs were necessary, an abbreviated cycle was desirable. Current emission data on the two locomotive engines was analyzed and a "3-mode cycle" was found to be representative. A comparison of the two cycles is given in Table 4 and a comparison of typical results obtained with the two cycles on the EMD and GE engines is shown in Table 5. Reference 1 gives a detailed description of the procedure for calculating the weighted composite brake-specific emissions.

Table 4. Comparison of Full-Mode vs. 3-Mode Duty Cycles

Notch Position	Full-Mode Weighting Factor	3-Mode Cycle Weighting Factor
8	14	25
7	3	---
6	3	---
5	4	25
4	4	---
3	3	---
2	5	---
1	5	---
Idle	27.5	50
Low Idle	27.5	---
Dynamic Brake #1	2	---
Dynamic Brake #2	2	---
Total	**100 %**	**100 %**

Table 5. Full-Mode versus 3-Mode Composite Baseline Emissions

	weighted kW	Part. g/hp-hr	HC g/hp-hr	CO g/hp-hr	NO_X g/hp-hr
EMD 12-645E3B					
Full-Mode	432	0.28	0.33	0.80	11.7
3-Mode	661	0.27	0.29	0.60	11.3
GE 12-7FDL					
Full-Mode	436	0.26	0.60	2.24	10.7
3-Mode	703	0.22	0.41	1.88	10.8

For the EMD engine data given in Table 5, the 3-mode cycle gives reasonable results for PM and HC, but it tends to yield NO_X results that are about 3 percent lower than that calculated with the full-mode cycle. For the GE engine, the NO_X results correlate closely while the PM results are about 15 percent lower and the HC results are about 30 percent lower with the 3-mode cycle. These differences in PM and HC are misleading however, because these emissions are very low whether calculated with the full-mode or 3-mode method. Calculated CO emissions tend to be 13 to 25 percent lower on both engines using the 3-mode duty cycle, however this is considered acceptable since CO emissions are very low on these engines when compared to other mobile sources.

In the interest of maximizing the number of control techniques which could be evaluated, it was decided to use the 3-mode cycle for all development testing. If a more accurate correlation with the full-mode cycle is desired, a detailed study of emission results may suggest a slight adjustment of the weighting factors.

EXPERIMENTAL RESULTS

Presented below is a brief description of the engine and fuel modifications that were performed for this test program. A more detailed description of the hardware modifications is available [17,18]. The highlights of the test results are summarized in the following sections, with comprehensive tabulated data given in Appendix A for each test configuration. Emissions data for each individual test point are also available [18].

Injection Timing Effects

The effect of retarded timing was investigated by varying the static injection timing over the range of adjustment. In the case of the EMD engine, this was from standard (TDC) to 6°ATDC. For the GE engine, adjustments were made from the standard injection timing of 24°BTDC to 12°BTDC.

The effects of fuel injection timing on NO_X and particulate for the EMD engine over the composite 3-mode cycle results are shown in Figure 2. Note that these test results were obtained with baseline diesel fuel and an unmodified aftercooler system. With a 4° retard, NO_X was reduced from a baseline level of 10.5 g/hp-hr to 8.9 g/hp-hr. Particulate, however, increased from 0.15 g/hp-hr to 0.24 g/hp-hr, and fuel consumption increased 1.3 percent for the EMD.

Figure 2.

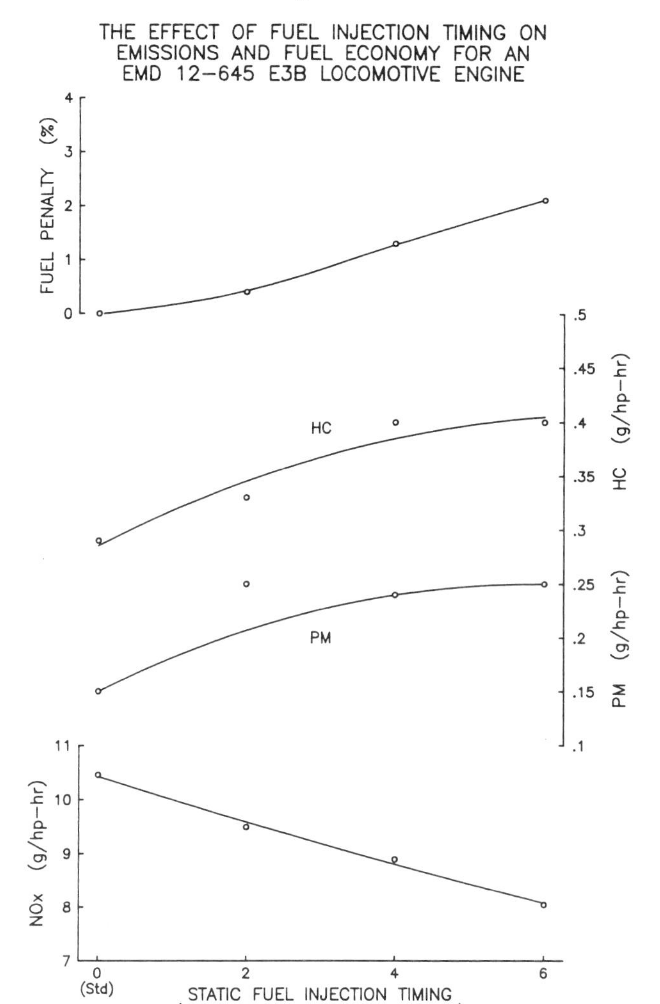

For the GE engine, fuel injection timing effects using baseline fuel are shown in Figure 3. A 4° timing retard decreased NO_X from a baseline condition of 11.2 g/hp-hr to 9.2 g/hp-hr, while particulate increased from 0.19 g/hp-hr to 0.26 g/hp-hr, and fuel consumption increased 1.2 percent.

Enhanced Aftercooling Effects

For investigations on the effect of enhanced aftercooling, a separate aftercooler water system was fabricated. The existing engine mounted aftercooler was utilized; however, engine cooling water no longer was used. Instead, utility water was the cooling medium, and the air temperature out of the aftercooler was varied by changing the water flow rate with a modulating valve. At Notch 8 and Notch 5, the reduced aftercooler temperatures were maintained, but at idle, the cooling water flow was shut off. Heating the intake air would have been required to maintain the target idle air temperature, and this would not have correctly simulated the field application. For all evaluations at Notch 8, Notch 5 and idle, three emission runs were made and the results averaged.

Figure 3.

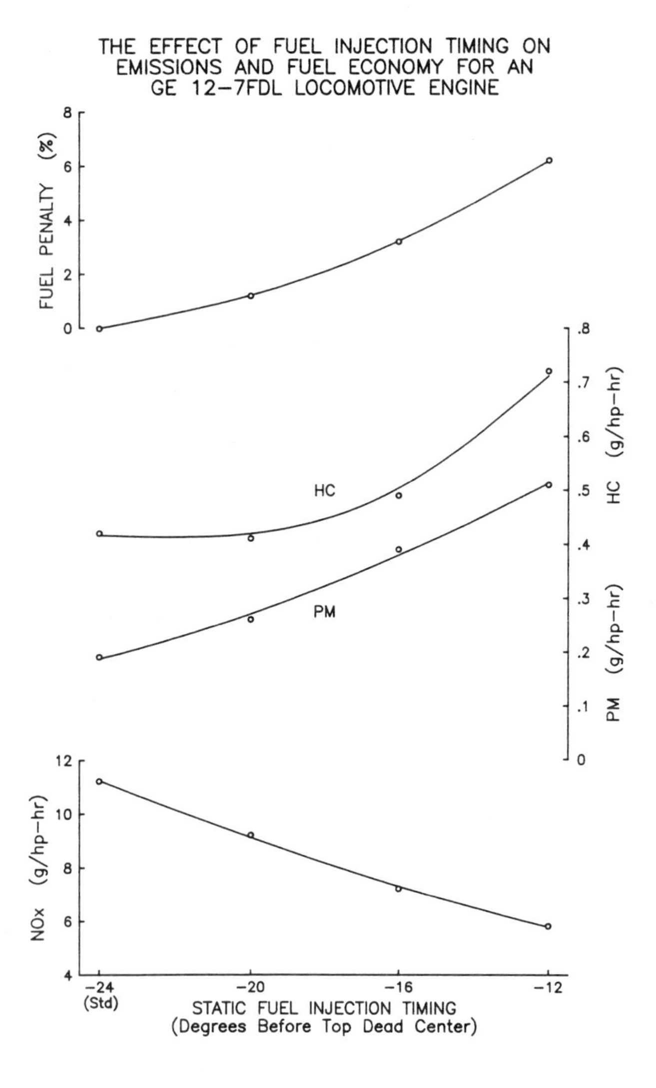

Enhanced intake air aftercooling experiments were performed on both the EMD and the GE engines with 4° timing retard. Results for the EMD engine are shown in Figure 4. A slight reduction in NO_X emissions was demonstrated, with NO_X decreasing from 8.9 g/hp-hr (with a typical boost air temperature after the aftercooler of 99°C) to 8.1 g/hp-hr at 91°C (a 9 percent reduction). Further reductions in boost air temperature demonstrated only nominal reductions in NO_X emissions. Three-mode cycle particulates, CO, HC, and fuel efficiency were mostly unaffected by boost air temperature changes.

Results for the GE engine are shown in Figure 5. The expected reduction in NO_X emissions was not as pronounced as observed for the EMD engine. NO_X decreased from 9.2 g/hp-hr at a typical boost air temperature (after the aftercooler) of 85°C, to 8.7 g/hp-hr at 77°C (a 5 percent reduction). Further reductions in boost air temperature also demonstrated slight reductions in NO_X emissions. Like the EMD engine, particulates, CO, HC, and fuel efficiency were mostly unaffected by boost air temperature changes, until boost air temperatures were forced below 68°C, where increases in CO and HC were observed.

Low-Sulfur, Low-Aromatic Fuel Effects

Low-sulfur, low-aromatic diesel fuel was tested at standard timing with unmodified aftercooling, and with 4° injection retard using enhanced boost air aftercooling for both engines. The results of these tests are given in Table 6. The data indicate that at the standard injection timing, the low-sulfur, low-aromatic fuel did not significantly affect particulate emissions for either engine. This result is similar to results obtained in original baseline emission test work performed in 1989 using a very low sulfur fuel (0.01 wt. percent) [1]. However, when timing was retarded 4°, and enhanced aftercooling was in use, the low-sulfur, low-aromatic fuel was effective in reducing particulate emissions for both the EMD and GE engines.

The fuel penalty given in Table 6 represents any observed change in the engine brake specific fuel consumption, and the effect of the energy per unit volume difference of the low-sulfur, low-aromatic fuel compared to the baseline diesel fuel. Referring to Table 3, this difference is significant, with the low-sulfur, low aromatic fuel having 3.5 percent less energy per gallon than the baseline fuel.

Figure 4.

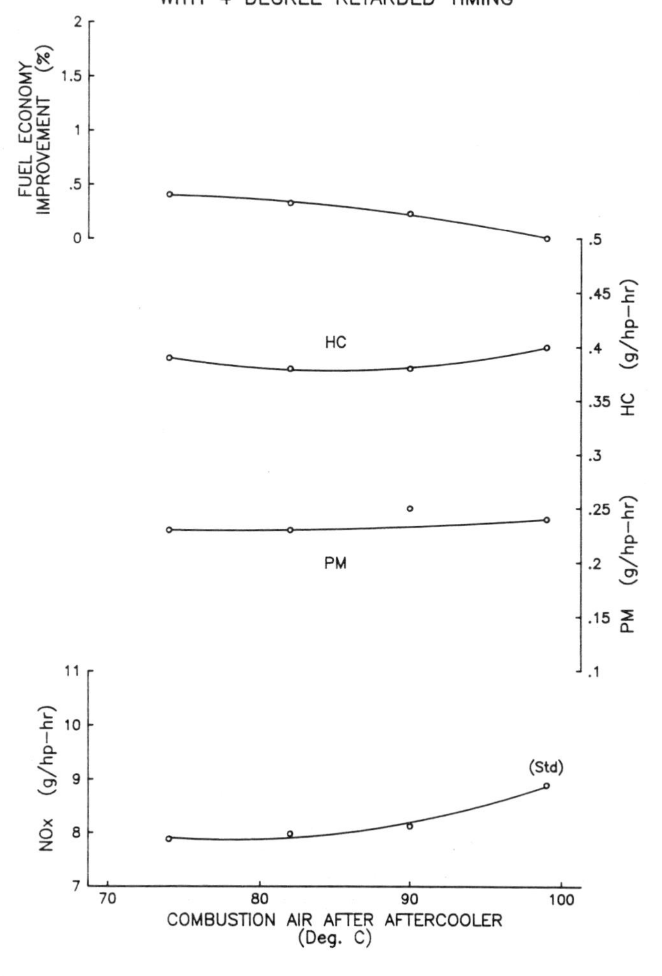

Figure 5.

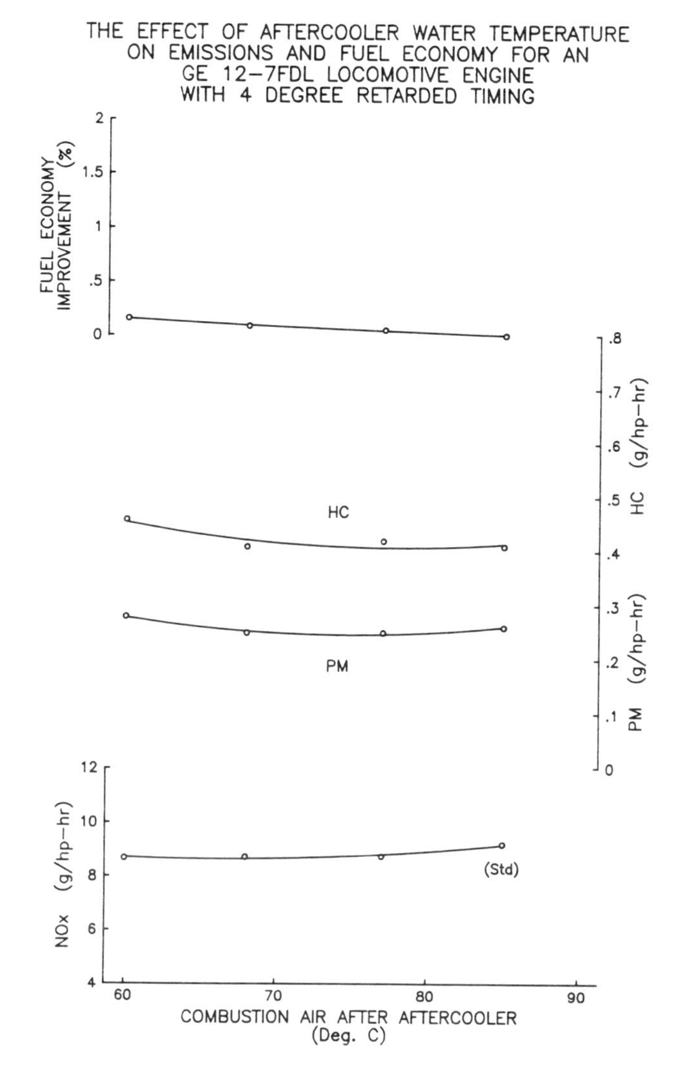

DISCUSSION / SUMMARY

One of the locomotive exhaust emissions reduction recommendations presented in the CARB Locomotive Emissions Study [2] was to use the combined implementation of a 4° injection timing retard and the use of a low-sulfur, high cetane diesel fuel for yard and switcher locomotives. The experimental efforts described above were an attempt to quantify the emissions and fuel economy changes associated with performing these recommended modifications. Based on the laboratory testing performed in this work, the following results can be expected by combining 4° injection timing retard; the use of a low-sulfur, low aromatic (and high cetane); and enhanced aftercooling:

EMD 645 E3B:

- NO_X reduced 31 percent.
- Particulate increased 27 percent.
- CO increased 26 percent.
- HC increased 14 percent.
- Fuel consumption increased 4.3 percent.

GE 7FDL

- NO_X reduced 32 percent.
- Particulate increased 5 percent.
- CO increased 9 percent.
- HC decreased 12 percent.
- Fuel consumption increased 4.2 percent.

Inspection of these data shows that significant changes to emissions and fuel consumption can be expected by implementing the changes recommended in the CARB study. One important finding is that the two major locomotive engine types respond differently to the modifications. Although NO_X was reduced, other emissions increased. Therefore, emissions and fuel penalty trade-offs must be considered before proceeding with modifications for lower NO_X emissions alone.

FUTURE WORK

A standardized test procedure for measuring locomotive engine emissions does not exist. Recognizing that a formal duty-cycle based test procedure for measuring gaseous and particulate locomotive emissions must be established, the AAR has initiated an industry working group to establish these procedures. The working group is made up of representatives from EMD, GE, Caterpillar, AAR, and SwRI. It is anticipated that the standardized test procedure that evolves will be recognized by the locomotive engine manufacturers and will be adopted as an AAR Recommended Practice. This will ensure that future locomotive emissions are measured and reported in a consistent manner.

During the 1991 AAR research program, additional efforts will be made toward reducing exhaust emissions from locomotive engines. Studies will include determining the applicability of existing NO_X correction factors for ambient air temperature and humidity; assessing the contribution of lubricating oil consumption to particulate emissions; further evaluating fuel effects on emissions; and evaluating various engine hardware configurations for emissions effects.

ACKNOWLEDGEMENTS

Thc AAR Phase XII Diesel Fuel Specification and Locomotive Improvement Program was performed at Southwest Research Institute for the Association of American Railroads. The AAR Project Technical Monitor was Mr. G. Richard Cataldi. Guidance for the Program was provided by the AAR Locomotive Efficiency Review Committee and the AAR Energy and Locomotive Program Steering Committee comprised of individuals from the member railroads of the AAR and locomotive manufacturers.

REFERENCES

1. Fritz, S.G. and G.R. Cataldi, "Gaseous and Particulate Emissions From Diesel Locomotive Engines," New Technology in Large Bore Engine, ASME ICE-Vol 13, B. Chrisman- Editor, pp. 63-72, October 1990.

2. "Locomotive Emissions Study," by Booz-Allen & Hamilton, Final Report for the California Air Resources Board, January 1991.

Table 6. The Effect of Low-Sulfur, Low-Aromatic Fuel on Locomotive Engine Exhaust Emissions

	PM (g/hp-hr)	HC (g/hp-hr)	CO (g/hp-hr)	NO_X (g/hp-hr)	Fuel Penalty (%)
EMD 12-645 E3B					
Baseline Fuel, Std. Timing	0.15	0.29	0.81	10.5	---
Low Aromatic Fuel, Std. Timing	0.17	0.33	0.83	10.4	3.5[a]
Baseline Fuel, 4° Retard, Enhanced Aftercooling	0.23	0.38	0.92	8.0	0.4[b]
Low Aromatic Fuel, 4° Retard, Enhanced Aftercooling	0.19	0.33	1.02	7.2	4.3[c]
GE 12-7FDL					
Baseline Fuel, Std. Timing	0.19	0.42	1.98	11.2	---
Low Aromatic Fuel, Std. Timing	0.18	0.36	1.87	9.9	3.5[a]
Baseline Fuel, 4° Retard, Enhanced Aftercooling	0.26	0.41	1.91	9.2	1.2[b]
Low Aromatic Fuel, 4° Retard, Enhanced Aftercooling	0.20	0.36	2.04	7.6	4.2[c]

Notes: All emissions are three-mode weighted composite.
a - Reflects btu/gallon difference only.
b - Reflects bsfc difference only.
c - Reflects bsfc and btu/gallon difference.

3. "Management Plan For Nitrogen Oxides (NO_X) and Volatile Organic Compounds (VOCs) - Phase I), Canadian Council of Ministers of the Environment, CCME-EPC/TRE-31E, November 1990.

4. "SP-AAR Program to Develop Certification Procedures With Respect to Visible Emissions from New and Out-Shopped Locomotives," SP Final Report, August 1973.

5. Hoffman, J.G., K.J. Springer, and T.A. Tennyson, "Four Cycle Diesel Electric Locomotive Exhaust Emissions: A Field Study," ASME Paper No. 75-DGP-10, April 1975.

6. Hare, C.T. and K.J. Springer, "Exhaust Emissions From Uncontrolled Vehicles and Related Equipment Using Internal Combustion Engines: Part 1 - Locomotive Diesel Engines and Their Marine Counterparts," SwRI Report No. 11-2869-001, 1972.

7. Storment, J.O. and K.J. Springer, "Assessment of Control Techniques For Reducing Emissions From Locomotive Engines," SwRI Report No. AR-844, April 1973.

8. Hare, C.T., K.J. Springer, and T.A. Huls, "Locomotive Exhaust Emissions and Their Impact," ASME Paper No. 74-DGP-3, May 1974.

9. Storment, J.O., K.J. Springer, and K.M. Hergenrother, "NO_X Studies With EMD 2-567 Diesel Engine," ASME Paper No. 74-DGP-14, May 1974.

10. Springer, K.J. and O.J. Davis, "Studies of NO_X Emissions From A Turbocharged Two-Stroke Cycle Diesel Engine," SwRI Report No. 11-2869-003, October 1975.

11. Conlon, P.C., "Exhaust Emission Testing of In-Service Diesel-Electric Locomotives - 1981 to 1983," AAR Report No. R-688, March 1988.

12. Baker, Q.A., et al, "Alternative Fuels for Medium-Speed Diesel Engines Program, Fourth Research Phase Final Report," SwRI Report No. 03-7446-001, AAR Report No. R-569, February 1984.

13. Wakenell, J.F., et al, "Alternative Fuels for Medium-Speed Diesel Engines Program, Fifth Research Phase Final Report," SwRI Report No. 03-7924, AAR Report No. R-602, April 1985.

14. Wakenell, J.F., et al, "Alternative Fuels for Medium-Speed Diesel Engines Program, Sixth Research Phase Final Report," SwRI Report No. 03-8469, AAR Report No. R-615, October 1986.

15. Wakenell, J.F., et al, "Diesel Fuel Specification and Locomotive Improvement Program, Eight Research Phase Final Report," SwRI Report No. 03-1542, AAR Report No. R-697, December 1987.

16. Fritz, S.G., et al, "Diesel Fuel Specification and Locomotive Improvement Program, Ninth Research Phase Final Report," SwRI Report No. 03-2082, AAR Report No. R-731, August 1989.

17. Fritz, S.G., et al, "Diesel Fuel Specification and Locomotive Improvement Program, Tenth Research Phase Final Report," SwRI Report No. 03-2695, AAR Report No. R-771, December 1989.

18. Markworth, V.O., et al, "Diesel Fuel Specification and Locomotive Improvement Program, Eleventh Research Phase Final Report," SwRI Draft Final Report No. 03-3324, June 1991.

Appendix A
Locomotive Engine Emissions Test Summary

GE 12-7FDL Emissions Data								
	Fuel Type	Timing Retard (Deg)	Boost Air Temp. (°C)	BSPM (g/hp-hr)	BSHC (g/hp-hr)	BSCO (g/hp-hr)	$BSNO_x$ (g/hp-hr)	Power/Fuel Loss (%)
Baseline	Baseline	0	85	0.19	0.42	2.0	11.2	0.0
4° Retard	Baseline	4	85	0.26	0.41	1.9	9.2	1.2[b]
8° Retard	Baseline	8	85	0.39	0.49	2.2	7.2	3.2[b]
12° Retard	Baseline	12	85	0.51	0.72	2.6	5.8	6.2[b]
Baseline - Low Aromatic Fuel	Low Aromatic	0	85	0.18	0.36	1.9	9.9	3.5[a]
Enhanced Aftercooling (4° Retard)								
85°C - Standard Condition	Baseline	4	85	0.26	0.41	1.9	9.2	0.0[b]
77°C	Baseline	4	77	0.25	0.42	1.9	8.7	0.1[b]
68°C	Baseline	4	68	0.25	0.41	2.0	8.7	0.2[b]
60°C	Baseline	4	60	0.28	0.46	2.0	8.7	0.3[b]
4° Retard, Low Aromatic, Enhanced Aftercooling	Low Aromatic	4	60	0.20	0.36	2.0	7.6	4.2[a]

Notes: a - Reflects bsfc changes and fuel btu/gallon differences.
b - Reflects only bsfc difference.

EMD 12-645E3B Emissions Data								
	Fuel Type	Timing Retard (Deg)	Boost Air Temp. (°C)	BSPM (g/hp-hr)	BSHC (g/hp-hr)	BSCO (g/hp-hr)	$BSNO_x$ (g/hp-hr)	Power/Fuel Loss (%)
Baseline	Baseline	0	99	0.15	0.29	0.81	10.5	0.0
2° Retard	Baseline	2	99	0.25	0.33	0.77	9.5	0.4[b]
4° Retard	Baseline	4	99	0.24	0.40	0.85	8.9	1.3[b]
6° Retard	Baseline	6	99	0.25	0.40	0.95	8.0	2.1[b]
Baseline - Low Aromatic Fuel	Low Aromatic	0	99	0.17	0.33	0.83	10.4	3.5[a]
Enhanced Aftercooling (4° Retard)								
99°C - Standard Condition	Baseline	4	99	0.24	0.40	0.85	8.9	0.0[b]
91°C	Baseline	4	91	0.25	0.38	0.95	8.1	0.2[b]
82°C	Baseline	4	82	0.23	0.38	0.92	8.0	0.3[b]
74°C	Baseline	4	74	0.23	0.39	0.93	7.9	0.4[b]
4° Retard, Low Aromatic, Enhanced Aftercooling	Low Aromatic	4	82	0.19	0.33	1.0	7.2	4.3[a]

Notes: a - Reflects bsfc changes and btu/gallon differences.
b - Reflects only bsfc difference.

ICE-Vol. 15, Fuels, Controls, and Aftertreatment
For Low Emissions Engines
ASME 1991

ELECTRICALLY-HEATED CATALYSTS FOR COLD-START EMISSION CONTROL ON GASOLINE- AND METHANOL-FUELED VEHICLES

Martin J. Heimrich
Department of Emissions Research
Southwest Research Institute
San Antonio, Texas

Steve Albu
State of California
Air Resources Board
El Monte, California

Manjit Ahuja
State of California Air Resources Board
Sacramento, California

ABSTRACT

Cold-start emissions from current technology vehicles equipped with catalytic converters can account for over 80 percent of the emissions produced during the Federal Test Procedure (FTP). Excessive pollutants can be emitted for a period of one to two minutes following cold engine starting, partially because the catalyst has not reached an efficient operating temperature. Electrically-heated catalysts, which are heated prior to engine starting, have been identified as a potential strategy for controlling cold-start emissions. This paper summarizes the emission results of three gasoline-fueled and three methanol-fueled vehicles equipped with electrically-heated catalyst systems. Results from these vehicles demonstrate that heated catalyst technology can provide FTP emission levels of non-methane organic gases (NMOG), carbon monoxide (CO), and oxides of nitrogen (NO_x) that show promise of meeting the Ultra-Low Emission Vehicle (ULEV) standards established by the California Air Resources Board.

NOMENCLATURE

ARB State of California Air Resources Board
CFM Cubic feet per minute
CFR Code of Federal Regulations
CO Carbon Monoxide
CO_2 Carbon Dioxide
CRC Coordinating Research Council
°C Degrees Celsius
EHC Electrically-Heated Catalyst
EPA United States Environmental Protection Agency
FFV Flexible Fuel Vehicle produced by Ford Motor Company
FTP EPA Federal Test Procedure
g/mi Grams per mile
HC Hydrocarbons
HCHO Formaldehyde
H_2O Water
L Liters
LEV Low Emission Vehicle
L/min Liters per minute
M85 Fuel consisting of 85 percent methanol and 15 percent gasoline
M90 Fuel consisting of 90 percent methanol and 10 percent gasoline
NMHC Non-Methane Hydrocarbons
NMOG Non-Methane Organic Gases
NO_x Oxides of Nitrogen
O_2 Oxygen
SAE Society of Automotive Engineers
sec Second
SwRI Southwest Research Institute
TLEV Transitional Low Emission Vehicle
ULEV Ultra-Low Emission Vehicle
VFV Variable Fuel Vehicle produced by General Motors Company
V-6 6-Cylinder Engine with "V" Configuration
V-8 8-Cylinder Engine with "V" Configuration
w/ with

INTRODUCTION

Cold-start emissions are recognized as the greatest contributor to Federal Test Procedure (FTP) emissions for both gasoline- and methanol-fueled vehicles. Seventy to eighty percent of FTP exhaust emissions from current-technology vehicles are typically emitted during the first minute of cold operation. This is because the catalyst is too cold to be active during this period. One strategy to control cold-start emissions is to electrically heat the catalyst prior to cold starting the engine. Other strategies that have been researched include cold-start emission collection, close-coupled catalysts (close to the exhaust manifold), advanced catalyst formulations, and latent heat storage devices. (1,2)* This paper will review the electrically-heated catalyst strategy for cold-start emission control and the California initiative to lower exhaust emission standards.

The State of California recently adopted new motor vehicle emission standards for 1994 and beyond. (3) These standards include limits for non-methane organic gases (NMOG), carbon monoxide (CO), oxides of nitrogen (NO_x) and formaldehyde (HCHO). Non-methane organic gases consist of all measurable reactive hydrocarbons (HC) and are defined as non-methane hydrocarbons (NMHC), aldehydes and ketones, and alcohols containing 12 or fewer carbon atoms. (4) Table 1 provides certification emission standards for designated Transitional Low Emission Vehicles (TLEVs), Low Emission Vehicles (LEVs) and Ultra-Low Emission Vehicles (ULEVs). Zero Emission Vehicles (ZEVs) are a fourth category and produce no emissions of any pollutant. These standards were adopted in September of 1990.

*Numbers in parentheses designate references at end of paper.

Table 1. California Exhaust Emission Standards for Light-Duty Vehicles

Vehicle Category	Exhaust Emissions, g/mi			
	NMOG	CO	NO_x	HCHO
Adopted for 1993	0.250	3.4	0.40	--
TLEV	0.125	3.4	0.40	0.015
LEV	0.075	3.4	0.20	0.015
ULEV	0.040	1.7	0.20	0.008
TLEV, LEV, ULEV Standards Adopted September 1990				

The California Air Resources Board has sponsored research on several electrically-heated catalyst applications at the Southwest Research Institute (SwRI). In two separate emission control studies, both gasoline- and methanol-fueled vehicles were modified by incorporating advanced catalyst technologies. One of the common objectives of these studies was to reduce the quantity of total organic gas emissions through the development of a total emission control system, without sacrificing control of other pollutants. The continued control of NO_x emissions was of special interest because of the NO_x contribution to smog formation.

A summary of the emission results from three gasoline-fueled vehicles and three methanol-fueled vehicles equipped with electrically-heated catalysts systems is presented. The gasoline-fueled vehicles were a 1986 Toyota Camry, a 1990 Buick LeSabre, and a 1990 Toyota Celica. The methanol-fueled vehicles selected for electrically-heated catalyst application studies were a hybrid 1981 Ford Escort equipped with a 1983 methanol-fueled engine, a 1988 Chevrolet Corsica VFV, and a 1989 Ford Crown Victoria FFV. Emissions data generated by these vehicles have been used to demonstrate the feasibility of recent California Low Emission Vehicle standards.

BACKGROUND

A cold-start is defined in the Code of Federal Regulations (CFR) as an engine start following a 12- to 36-hour continuous vehicle soak in a constant temperature environment of 20°C to 30°C. (5) Electrically-heated catalysts have been developed to assist conventional catalysts with cold-start emission control. An electrically-heated catalyst is an automotive exhaust catalyst with an electrically conductive support. Electric current from the vehicle battery directly heats the catalyst support (not the exhaust gas). Because the heated catalyst is hot prior to engine cranking, catalytic activity is possible and cold-start emissions can be controlled.

To initiate the catalyst heating sequence on today's demonstration vehicles, the ignition switch is placed in the "on" position. The driver then waits to start the engine while the catalyst is being heated. When the catalyst has reached a predetermined light-off temperature, the driver starts the engine. These heated catalyst systems modulate electrical power to the catalyst if the bed temperature drops below the light-off temperature. Additional electrical power is supplied to the catalyst for only a short time following the engine start. Typically, post-start heating continues for 10 to 50 seconds following a cold start and for less time following a hot start.

In an Internal Research study conducted by SwRI, a gasoline-fueled vehicle was equipped with an electrically-heated catalyst. This study produced the initial finding that heating alone did not significantly improve emission control over an unheated catalyst test. (1,6) With only electrical heating, cold-start catalyst activity was still impaired because of a lack of oxygen in the exhaust. In many vehicle fuel system calibrations, a cold engine is run fuel-rich to maintain driveability. Rich fuel-air ratios result in insufficient oxygen levels in the raw exhaust, limiting the oxidation of organic gases and CO.

To overcome this oxygen deficiency, secondary air was added for the conversion of organic gases and CO to carbon dioxide (CO_2) and water (H_2O). Pumps were used to inject air ahead of the heated catalyst following cold engine starting. On some vehicles, a short period of air injection following the hot start was beneficial. Emissions of NO_x cannot be controlled in the presence of excess oxygen. This lack of control occurs because oxygen-rich environments do not provide a suitable environment for the chemical reduction of NO_x. For this reason, short air injection periods are favored to maintain NO_x emission control. In the heated catalyst applications described in this paper, carefully defined air injection strategies were identified to minimize increases of NO_x emissions.

GASOLINE-FUELED VEHICLE HEATED CATALYST APPLICATIONS

Three gasoline-fueled vehicles were equipped with electrically-heated catalyst systems at SwRI. These vehicles were a 1986 Toyota Camry, a 1990 Buick LeSabre, and a 1990 Toyota Celica. A summary of these heated catalyst installations and emission improvements is presented for each vehicle.

1986 Toyota Camry

A gasoline-fueled 1986 Toyota Camry was obtained for an electrically-heated catalyst research study. (1,6) This vehicle was equipped with a 2.0 liter engine, electronic port fuel injection, and a three-way (only) catalytic converter. The original underbody stock catalyst was removed and replaced with an electrically-heated catalyst, a three-way formulation designed as a total replacement for the stock catalyst. At the start of the heated catalyst research, this vehicle had almost 35,400 kilometers (22,000 miles) on the odometer.

The electrically-heated catalyst used on this vehicle was composed of two separate sections. The larger downstream section was a metal substrate catalyst without heating capability while the smaller upstream section was a catalyzed metal substrate with the ability to be heated electrically. Figure 1 shows a schematic of this electrically-heated catalyst. The electrically-heated catalyst underwent 805 kilometers (500 miles) of service accumulation on the Camry prior to experimentation. A photograph of the actual electrically-heated catalyst unit is given in Figure 2.

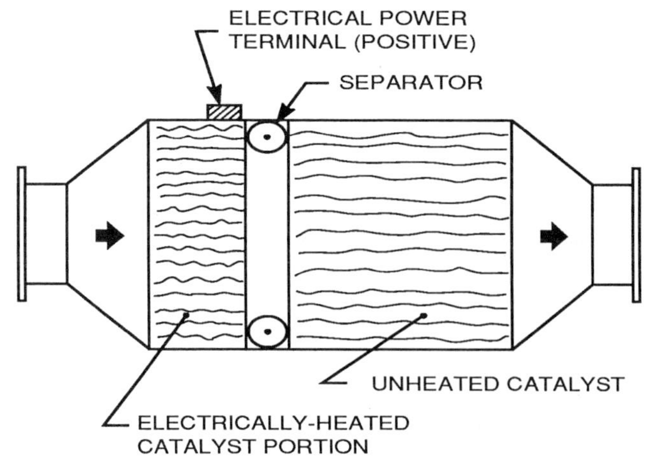

Figure 1. Schematic of Electrically-Heated Catalyst Used on Toyota Camry

Figure 2. Electrically-Heated Catalyst on Toyota Camry

The Toyota Camry underwent several cold-start FTP experiments in order to determine the optimal air injection flowrate and duration. Air was injected ahead of the heated catalyst at a rate of 140 liters per minute (5.0 cubic feet per minute) for the first 140 seconds of the cold-start portion of the FTP. An off-vehicle electrically-driven air pump was used for injecting air ahead of the heated catalyst. Standard catalyst heating times were established. Electrical heating was performed for 15 seconds before and for 30 seconds after the cold start. For hot starts, heating times were 5 seconds before and 10 seconds after hot cranking. Catalyst temperatures typically reached 500°C to 600°C prior to engine starting.

Federal Test Procedure emission tests were performed on the Camry with different combinations of catalyst heating and air injection. As described in Reference 6, optimal heated catalyst performance occurred with both catalyst preheating and secondary air injection. These emissions results are compared to no-heat-no-air baseline emissions in Table 2. Significant reductions in NMHC and CO were obtained with the air-injected heated catalyst. Emissions of NO_x increased only slightly because of the secondary air.

Table 2. Toyota Camry Heated Catalyst Emissions Compared to Baseline

Catalyst Configuration	FTP Emissions, g/mi		
	NMHC	CO	NO_x
Baseline	0.12	1.13	0.22
EHC w/Air	0.03	0.38	0.25

Air Injection: 140 L/min, 140 sec. for cold-start.
Baseline: Camet catalyst without heat nor air.

1990 Buick LeSabre

A new 1990 Buick LeSabre equipped with a three-way underbody catalyst and a port fuel injected 3.8 liter engine was outfitted with a heated catalyst system for the California Air Resources Board. (1,7) The electrically-heated catalyst was mounted just ahead of the underbody production catalyst, as shown in Figure 3. An electric air pump was mounted under the hood. An aftermarket battery with more energy capacity replaced the original battery. The heated catalyst power controller was mounted inside the vehicle.

Buick LeSabre

Figure 3. Electrically-Heated Catalyst Mounted in Front of Production Catalytic Converter

A series of developmental air injection experiments was performed to determine the optimal air injection strategy. Cold-start and hot-start air injection strategies were investigated using the laboratory air injection pump. After several FTP emission tests, an air injection flowrate of 300 liters per minute (10.7 cubic feet per minute) was determined to be optimum. Cold-start air injection began as the engine cranked and continued for 75 seconds. Hot-start air injection began as the engine cranked and continued for 30 seconds.

Electrical heating of the catalyst on the Buick LeSabre was regulated by the on-vehicle power controller. This controller was programmed to maintain the catalyst bed above a temperature of 370°C. A thermocouple inside the heated catalyst provides a temperature feedback to the power controller. When the catalyst underwent prestart heating, the power controller heated the catalyst until the catalyst reached the programmed temperature. At this time this engine was started. After the engine started, the catalyst was electrically heated (as needed) to maintain the programmed temperature. The catalyst temperature was monitored by the controller for a period of up to three minutes following a hot or cold start, after which no additional electrical heating occurred.

Electrically-heated catalyst emission test results are compared to the Buick stock catalyst emissions in Table 3. Emissions of NMHC and CO were reduced significantly while NO_x emissions were unchanged. Fuel economy decreased slightly due to additional (on-vehicle) battery recharging following the heating sequence.

Table 3. Buick LeSabre Heated Catalyst Emissions Compared to Stock

Catalyst Configuration	FTP Emissions, g/mi			Fuel Economy, mi/gal
	NMHC	CO	NO_x	
Stock (typical)	0.15	1.36	0.18	20.2
EHC w/air	0.02	0.25	0.18	19.7

Air Injection: 300 L/min; 75 sec. for cold-start, 30 sec. for hot-start.

1990 Toyota Celica

An electrically-heated catalyst system was also applied to a new 1990 Toyota Celica. (1,7) The Celica, shown in Figure 4, was factory-equipped with a close-coupled catalytic converter plus a relatively small underbody catalyst. There was no air injection for the original catalyst configuration. This vehicle provided an opportunity to study the emission control potential of an electrically-heated catalyst located downstream of a close-coupled catalyst. In this downstream location, the electrically-heated catalyst would be expected to operate at lower overall temperatures than if located first in the exhaust stream. Long-term durability for the heated catalyst in this downstream location may be enhanced in this cooler location.

Figure 4. 1990 Toyota Celica

The close-coupled production catalyst was found to absorb engine exhaust heat, which caused the downstream electrically-heated catalyst to warm up very slowly. Difficulties in maintaining the desired electrically-heated catalyst temperature caused extended post-start heating times and contributed to increased engine emissions (because of the increased battery recharging load on the alternator).

The application of electrically-heated catalyst technology to the Celica underwent several stages of optimization. Optimum air injection was determined to be 20 seconds following the cold-start and zero (no air injection) following the hot start. Post-start catalyst electrical heating was limited to 20 seconds in order to minimize battery discharge and subsequent recharging (energy replacement).

Toyota Celica emissions are given in Table 4 and show improved control of NMHC and CO, similar to other heated catalyst applications. For this application, the reduction in FTP NO_X emissions can be attributed to a relatively large heated catalyst volume. (1,7) The reduced post-start heating time contributed to a minimal fuel economy penalty, as compared to the extended heating times previously used. An in-depth analysis of this heated catalyst application is given in Reference 7.

Table 4. Toyota Celica Heated Catalyst Emissions Compared to Stock

Catalyst Configuration	FTP Emissions, g/mi			Fuel Economy, mi/gal
	NMHC	CO	NO_X	
Stock (typical)	0.09	0.68	0.08	25.6
EHC w/air	0.03	0.40	0.05	25.5
Air Injection: 300 L/min; 20 sec. for cold-start, no air for hot-start. Post-start heating limited to 20 sec., no hot-start heating.				

METHANOL-FUELED VEHICLE HEATED CATALYST APPLICATIONS

Three methanol-fueled vehicles were equipped with electrically-heated catalyst systems. These vehicles were a 1981 Ford Escort (with a 1983 engine), a 1988 Chevrolet Corsica VFV, and a 1989 Ford Crown Victoria FFV. The Escort and Crown Victoria used air injection strategies that were incorporated in their original emission control system. The Corsica used the off-vehicle (laboratory) air injection pump for emissions testing. A summary of these heated catalyst applications is presented for each vehicle.

1981 Ford Escort

A 1981 Ford Escort was used to demonstrate electrically-heated catalyst feasibility as part of an investigation into formaldehyde emission control for methanol-fueled vehicles. (8) The Escort had a 1.6 liter high output methanol-fueled engine (1983 version). This engine was equipped with a carburetor and calibrated to run on M90 fuel (90 percent methanol, 10 percent gasoline).

Four manufacturer-supplied catalyst formulations were evaluated for formaldehyde emission control for the California Air Resources Board. Three of the four catalysts screened were conventional (non-electrically heated) catalysts. The electrically-heated catalyst was similar to the one used on the 1986 Toyota Camry. The catalyzed, heated metal substrate was located upstream of an unheated metal substrate catalyst within the same container. All four sample catalysts (each with different formulations) were designed and tested as a total underbody catalyst replacement. These catalysts represented current methanol catalyst technology.

For this discussion, two of the catalysts evaluated were selected for comparison. One was the electrically-heated catalyst. The other was a conventional three-way plus oxidation catalyst with air injection between the pieces. This catalyst was selected because it demonstrated the best overall performance of the conventional catalysts.

Air injection and catalyst heating strategies were defined for the Escort. Cold-start secondary air was obtained using the air injection system from the pre-conversion (gasoline) 1981 Escort. Secondary air was delivered to the exhaust manifold (upstream of the electrically-heated catalyst). The originally-equipped on-vehicle air injection system was not optimized for this application. Fixed catalyst heating times were established for the Escort. For cold-starts, the catalyst was heated for 15 seconds prior to engine starting. There was no hot start catalyst pre-heating. (8)

Emission tests were run on the Ford Escort with the methanol catalysts. Methanol and formaldehyde emissions were measured and used to calculate NMOG. A comparison between the electrically-heated catalyst and the baseline FTP emission tests is given in Table 5. Formaldehyde, methanol, gasoline-derived hydrocarbons, and CO are primarily emitted during cold-start operation. (8) Generally, the electrically-heated catalyst was able to control formaldehyde, methanol, and CO emissions during the cold-start better than the other experimental catalysts used in this program, indicating that catalyst preheating and air injection made a significant improvement in cold-start emission control. This vehicle, however, still exhibited relatively high emissions of NMOG, probably because of its fuel system configuration and a limited heated catalyst optimization.

Table 5. Methanol Ford Escort Heated Catalyst Emissions Compared to Baseline

Catalyst Configuration	Fuel	FTP Emissions, g/mi			
		NMOG	CO	NO_x	HCHO
Baseline	M90	0.44	2.12	0.38	0.0093
EHC w/Air	M90	0.10	1.85	0.46	0.0075

Air Injection: Escort production air injection strategy.
NMOG = Gasoline-derived hydrocarbons + methanol + formaldehyde - methane.
NMOG calculation uses stock catalyst FTP emissions of 0.04 g/mi methane.
Each value is the average of two emission measurements.

1988 Chevrolet Corsica VFV

A 1988 Chevrolet Corsica VFV was selected for study in a formaldehyde emission control program by the California Air Resources Board. (8) This variable fuel vehicle (VFV) could be run on any mixture of M85 fuel and gasoline. The Corsica was originally equipped with a 2.8 liter V-6 engine, port fuel injection, and an underbody three-way catalyst. The Corsica odometer displayed 3200 km (2,000 miles) at the time of the heated catalyst application.

The Corsica VFV was equipped with a laboratory electrically-heated catalyst at SwRI. This catalyst was similar to the one installed on the LeSabre and Celica, except that an off-vehicle battery and air pump were used. Corsica emissions with the production three-way catalyst are compared to electrically-heated catalyst emissions in Table 6. Both emissions of NMOG and CO were reduced significantly. Even NO_x emissions were reduced moderately. Cold-start methanol and formaldehyde emission reductions contributed to lower FTP emissions with the heated catalyst and air injection. (8) Fuel economy figures were not compared on the Corsica because an off-vehicle battery (with no recharging) was used to preheat the catalyst.

Table 6. Chevrolet Corsica VFV Heated Catalyst Emissions Compared to Stock

Catalyst Configuration	Fuel	FTP Emissions, g/mi			
		NMOG	CO	NO_x	HCHO
Stock	M85	0.128	1.75	0.25	0.0213
EHC w/air	M85	0.03	0.40	0.17	0.0036

Air Injection: 100 sec for cold-start, 100 sec. for hot-start.
NMOG = gasoline-derived hydrocarbons + methanol + formaldehyde - methane.

1989 Ford Crown Victoria FFV

A 1989 Ford Crown Victoria FFV was selected to study emissions from advanced technology methanol-fueled vehicles. (8) This flexible fuel vehicle (FFV), was capable of running on any mixture of gasoline and M85 fuel. The Crown Victoria was originally equipped with a close-coupled three-way catalyst and an underbody oxidation catalyst in each exhaust bank of a V-8 (5.0 liter) engine (for a total of four catalysts; two close-coupled and two underbody).

The Crown Victoria was modified by SwRI with an electrically-heated catalyst system for laboratory emission demonstration tests. Electrically-heated catalysts were placed in each exhaust bank upstream of the original underbody catalysts. Cold-start air injection was provided by the original-equipment secondary air pump (designed for the stock catalyst application). This air injection pump provided air ahead of the heated catalyst during cold-start operation and directly to the downstream oxidation catalyst afterwards.

Emissions from both the heated and stock catalyst configurations, both running on M85 fuel, are given in Table 7. Significant improvements in NMOG and CO emissions were obtained with the heated catalyst system, however, NO_x emissions were similar to the baseline. Methanol emissions were reduced to undetectable levels with the heated catalyst configuration. (8)

Table 7. Ford Crown Victoria FFV Heated Catalyst Emissions Compared to Stock

Catalyst Configuration	Fuel	FTP Emissions, g/mi			
		NMOG	CO	NO_x	HCHO
Stock	M85	0.13	0.72	0.53	0.0090
EHC w/air	M85	0.01	0.23	0.52	0.0012

Air Injection: Crown Victoria production air injection strategy.
EHC located in each exhaust bank of V-8.
NMOG = gasoline-derived hydrocarbons + methanol + formaldehyde - methane.

SUMMARY

Electrically-heated catalyst applications on three gasoline-fueled vehicles and three methanol-fueled vehicles have been reviewed. Based on the results of these experiments, the electrically-heated catalyst with air injection has shown itself to be potentially capable of meeting future California emission standards. These heated catalyst systems, however, were unaged and were applied in a laboratory setting. Research into heated catalyst durability has begun (9). In addition, four of the vehicles converted with heated catalyst emission control systems described in this paper will undergo on-road service accumulation by the State of California. Follow-up studies on these vehicles and data from other roadworthy fleets should provide insight into more durable and reliable heated catalyst systems.

DISCLAIMER

The statements and conclusions in this report are those of the authors and not necessarily those of the California Air Resources Board. The mention of commercial products, their sources, or their use in connection with material reported herein is not to be construed as either an actual or implied endorsement of such products.

ACKNOWLEDGEMENTS

This paper is based on work performed by the Department of Emissions Research at Southwest Research Institute under ARB Contract No. A6-204-32, "Control of Benzene Emissions from Light-Duty Motor Vehicles," and ARB Contract No. A732-148, "Formaldehyde Emission Control Technology for Methanol-Fueled Vehicles." Initial air injection research was performed in the SwRI Internal Research Study, "Experimentation to Determine the Feasibility of Air Injection on an Electrically-Heated Catalyst for Reducing Cold-Start Benzene Emissions from Gasoline Vehicles," SwRI Project No. 08-9574. Use of the methanol-fueled Ford Escort was donated to SwRI and the ARB by the Coordinating Research

Council (CRC). The electrically-heated catalysts were provided by Camet Co. of Hiram, Ohio. The authors would like to recognize Juan Osborn and Sarah Santoro of the ARB and Lawrence Smith and Matthew Newkirk of SwRI for their contributions to the success of these programs.

REFERENCES

1. Heimrich, M.J., "Control of Benzene Emissions from Light-Duty Motor Vehicles," Final Report for the State of California Air Resources Board, Contract No. A6-204-32, Release Date: June 1991.

2. Schatz, O., "Cold-Start Improvements with a Heat Store," SAE Paper 910305, February 25 - March 1, 1991.

3. Staff Report, "Proposed Regulations for Low-Emission Vehicles and Clean Fuels," prepared by the State of California Air Resources Board, Mobile Source and Stationary Source Divisions, Release Date: August 13, 1990.

4. Technical Support Document for "A Proposal to Amend Regulations Regarding Exhaust Emission Standards and Test Procedures for Passenger Cars, Light-Duty Trucks, and Medium-Duty Vehicles for the Control of Criteria Pollutant and Toxic Air Contaminant Emissions," by the State of California Air Resources Board, Mobile Source Division, Draft, April 23, 1990.

5. Code of Federal Regulations, Title 40, Chapter 1, Part 86, Subpart B, Sections applicable to light-duty vehicles.

6. Heimrich, M.J., "Air Injection to an Electrically-Heated Catalyst for Reducing Cold-Start Benzene Emissions from Gasoline Vehicles," SAE Paper 902115, October 22-25, 1990.

7. Heimrich, M.J., Albu, S., and Osborn, J., "Electrically-Heated Catalyst System Conversions on Two Current-Technology Vehicles," SAE Paper 910612, February 25-March 1, 1991.

8. Current SwRI program, "Formaldehyde Emission Control Technology for Methanol-Fueled Vehicles," conducted by the Department of Emissions Research for the California Air Resources Board, Contract No. A732-148, SwRI Project 08-2346, 1991.

9. Whittenberger, W.A., Kubsh, J.E., "Electrically Heated Metal Substrate Durability," SAE Paper 910613, February 25-March 1, 1991.

ICE-Vol. 15, Fuels, Controls, and Aftertreatment
For Low Emissions Engines
ASME 1991

BENCH WEAR TESTING OF ENGINE POWER CYLINDER COMPONENTS

Donald J. Patterson
University of Michigan
Ann Arbor, Michigan

Stephen H. Hill
Sealed Power Technologies
Limited Partnership
Muskegon, Michigan

Simon C. Tung
General Motors Research Laboratories
Warren, Michigan

ABSTRACT

A need exists for an accurate and repeatable friction and wear bench test for engine power cylinder components that more closely relates to engine test results. Current research and development includes investigation of new engine designs, materials, coatings and surface treatments for reduced weight, longer life, higher operating temperatures, and reduced friction. Alternative fuels being examined include alcohols and gaseous fuels, as well as reformulated gasolines and distillate fuels. Concurrently, new lubricants are being formulated for the new engine and fuel combinations. Because of the enormous cost and time of developing commercial engine, fuel and lubricant combinations by means of engine testing alone, much interest is being focused on more representative and repeatable bench tests.

This paper examines some known bench testers employing either rotary or reciprocating motion for evaluating the friction, wear, and durability of material couples. Information is presented on experience and practice with one rotary (Falex type) and two reciprocating testers (Cameron-Plint and a new design, the EMA-LS9). Some correlation with engine data is given.

I INTRODUCTION

Engine, fuel and lubricant engineers are currently faced with a bewildering set of potential changes to the environment in which their products must successfully perform. Current research and development on power cylinder components includes new designs, materials, coatings and surface treatments. The goals are reduced weight, longer life, higher operating temperatures, and reduced friction. In addition, alternative fuels such as alcohols and gaseous compounds, as well as reformulated gasolines and distillate fuels are being examined. Concurrently, new lubricants are evolving for the new engine and fuel combinations.

The resulting systems must meet ever increasing environmental and fuel economy constraints. Because of the enormous cost and time of developing commercial engine, fuel and lubricant combinations by means of engine testing alone, much interest is being focused on more representative and repeatable bench tests.

Wear, scuff and friction testing of power cylinder components is commonly conducted in two stages: bench tests for rough screening and engine tests for final selection. The bench test offers a rapid and low cost means of comparison. However, since the engine environment is not completely simulated, engine tests will always be required for final verification. Of course, it is desirable to devise a bench test which more closely simulates the most severe wear conditions in the engine to the greatest extent possible.

The objective of this paper is to examine bench testers which have been used in recent years within the authors' organizations to screen ring and liner material couples. Information is presented on experience and practice with one rotary (Falex type) and two reciprocating testers (Cameron-Plint and a new design, the EMA-LS9). Statistical correlation with engine data is given for two of the testers.

II ROTARY TEST METHODS

Several configurations are in common use for wear testers using rotary motion. Two standard devices are the pin-on-disk (ASTM G99) and the block-on-ring (ASTM G77). Others include the ring-on-ring, ball-on-flat, four-ball, and thrust washer. The thrust washer test uses a rotating washer pressed against a stationary washer in a drill press like fixture.

For more than 20 years, a rotary block-on-ring tester (designated as LFW-1, Lubricant Friction and Wear Tester or Falex Tester, previously manufactured by DOW Corning, now made by Falex Corp.), Fig. 1, has been used in one author's laboratory to evaluate materials for piston rings, engine blocks and liners, valve guides, and gears; as well as coatings for piston rings. Scuff results have been obtained by incrementally increasing load until failure is detected audibly or by friction force measurement. Several procedures have been used for wear testing with this device. These involved variations in load, speed,duration, lubrication and temperature. Currently, the procedure delineated in Table 1 is used for virtually all testing on piston rings and liners. These conditions were selected to produce accelerated wear similar in nature to that found in engine tests. If this test is too severe for a given material, the load is reduced fourfold, and the test repeated with a new specimen. The coefficient of friction is determined at the beginning and end of each test from the weight applied and force transducer output. Block wear is most often evaluated by calculating the volume removed from the block wear scar area, although weight loss is an alternative. The ring wear is evaluated by weight loss or a subjective evaluation of visual appearance.

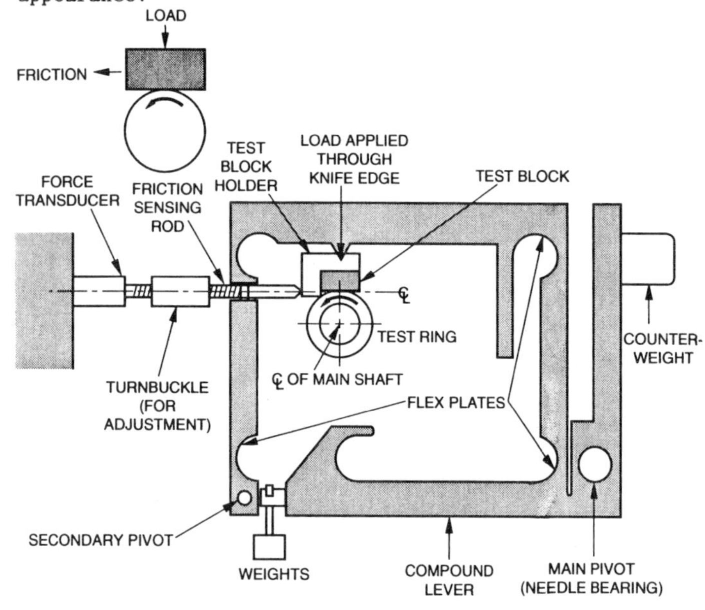

Fig. 1. Schematic of Falex type block-on-ring (LFW-1) wear tester.

Table 1
Test Procedure for Block-on-Ring (LFW-1) Tester

Block Size:	1.02 x 1.52 x 0.64 cm (optional ring segment mounted in block)
Surface Finish:	Block, finish with 600 grit paper Ring segment, use as received Ring, Finish with 220 grit paper
Speed:	197 rpm
Lubricant:	SAE 20W oil @ 90 - 93°C
Length of Test:	276,000 cycles (about 23 hrs)
Contact Load:	534 N
Load Application:	133 N initially, increased to 267 N @ 500 cycles, increased to 400 N @ 1000 cycles, increased to 534 N @ 1500 cycles and beyond

The test is initiated and brought to temperature under zero load. Initial readings are taken at 2000 cycles.

III RECIPROCATING TEST METHODS

Cameron-Plint. A modified Cameron-Plint High Frequency Friction Machine, Fig. 2, has been employed in another author's laboratory for bench test friction and wear evaluations of power cylinder materials. Previous examples of the use of the Cameron-Plint tester for simulated engine testing may be found in references [1-4]. For the ring/liner tests reported herein, a section of honed cylinder bore or liner is fixed in place, and a ring segment which is clamped in a modified head is moved across it with reciprocating motion in a horizontal plane, Fig. 3 (lower). The modified head incorporates a dowel pin to balance the ring segment, Fig. 3 (upper). The assembly is reciprocated by a scotch yoke mechanism driven by a variable speed motor. The ring is loaded against the liner by a spring balance. During the test, the tangential friction force is monitored by a piezo-electric transducer attached to the oscillating head. It is displayed continuously on an oscilloscope and digitized for storage and analysis.

A typical trace of friction versus time is shown in Fig. 4a. Note that the force reversal at the dead centers is not instantaneous, but occurs gradually as a result of elastic deformation. When the friction force reaches its peak, corresponding to zero velocity, energy is released from the moving ring. This causes the ring to accelerate as it attempts to "catch up" with the drive head. This "stick-slip" phenomena is consistent with friction characteristics at the top and bottom dead centers observed in engine tests [5]. Referring to Fig 4b, "T" is the period of one cycle and "t" is the interval between stroke end and point "a" where slip occurs. Point "b" is the mid-point of a single stroke where the friction force reaches a stabilized value, whereas "c" is the onset point of the return stroke when the direction of motion is reversed. At the instant when the ring finally comes to rest at the dead centers, the velocity of the drive head also reaches zero, and the deformation of the ring starts to diminish with the onset of another stroke. For this tester, a variety of test conditions, Table 2, has been selected to simulate several sliding regimes from boundary to hydrodynamic lubrication. Lubricant temperature of 150°C and test loads of 80N (about 900 psi cylinder pressure) to 120N represent heavy to severe engine loads. The speed range covers boundary through mixed and hydrodynamic lubrication regimes. The advantage of this tester is that a wide variety of test conditions and specimens can be rapidly evaluated. Friction force repeatability for three replicate tests is about $\pm$ 0.50 N (about 4-5%).

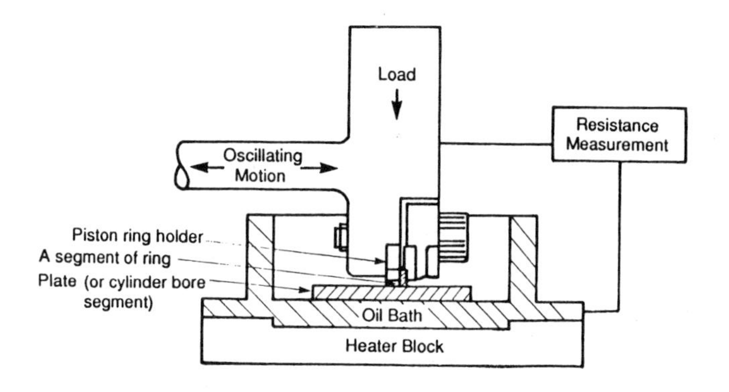

Fig. 2. Schematic of the Cameron-Plint tester.

Table 2
Test Details for Cameron-Plint
Friction and Wear Test Apparatus

Test Conditions

Temperature:	150oC
Speed:	10, 20, or 40 Hz (600, 1200, or 2400 rpm)
Stroke:	7.5 mm and 10 mm.
Load:	80 to 120 N (900 to 1350 psi cylinder pressure)
Duration:	60 or 120 hrs
Oil:	SAE 30 or 20W-20 engine oil

Measurements

1. Wear Scar and Surface Roughness
2. Wear Volume vs Sliding Time
3. Friction Coefficient vs Sliding Time
4. Surface Analysis by SEM, EPS, EDIX and AES.

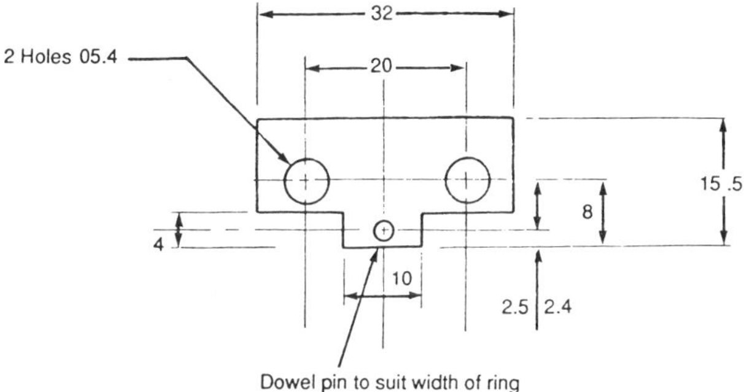

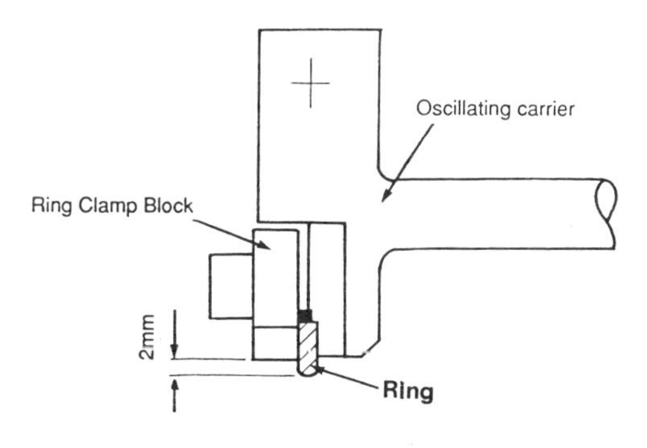

Fig. 3. Modified piston ring holder and arrangement for flat face ring section.

In our bench tests, we can evaluate different surface coatings for friction, wear resistance, compatibility, and adherence. It will be too time consuming and expensive to build engines for all of these coatings. This bench test can help us to sort different coating materials either on cylinder bores or piston rings. After sorting coating materials with a bench test, we will select good coating materials in engines. Engine test data can be compared with data generated from bench tests. The advantages of this bench test are that it can not only provide a fast approach to rank coating materials but it also can simulate engine conditions in a cost-effective way.

(a)

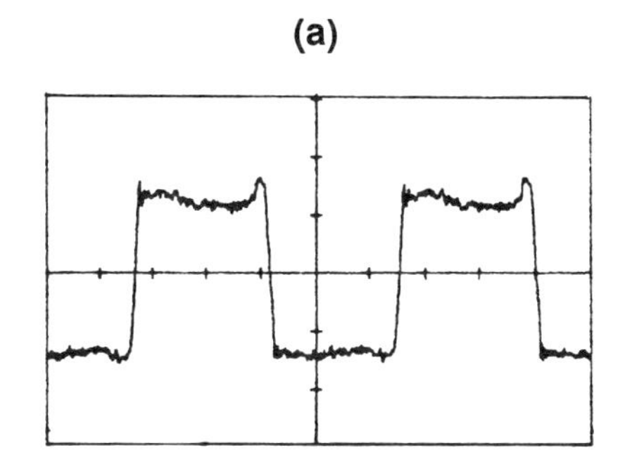

(b)

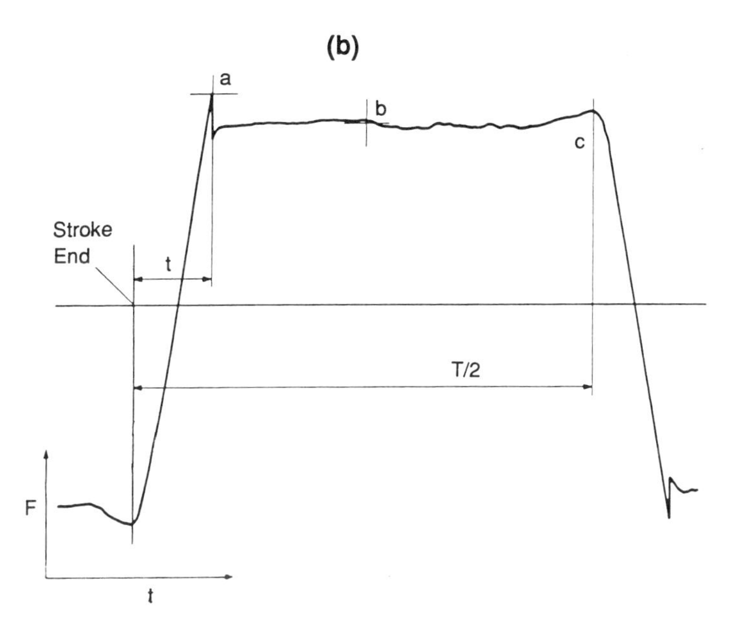

Fig. 4. Friction force as a function of stroke.
(a) Typical trace showing two cycles.
(b) Details of one stroke.

For the tests described herein, bore surface finish and plateau were varied, whereas the ring type was kept constant. For engine component tests, a ring from a smaller bore engine than that used for the liner was employed to eliminate edge contact loading. Tests with this machine may be conducted with a fully immersed or with a drip lubrication condition, the latter being used for the current tests. Friction force is reported as the root mean square (RMS) of the instantaneous friction force over one cycle. As an example, Fig. 5 shows RMS friction force as a function of time for simple plates with three different coatings. Differences in friction are evidenced by these tests. After testing, the wear scar and surface roughness of the tested pieces may be profiled using a computer-aided Surfanalyzer. Roughness profiles of production cylinder bores before and after 40 and 60 hours of sliding were analyzed and are presented as arithmetic mean average roughness (AA) and RMS roughness in Fig. 6. Results are shown for both length and width directions. Surface analyses such as SEM, XPS, EDIX, and Auger Electron Spectroscopy can determine the tribological characteristics and wear mechanisms of the tested components. In addition wear volume versus sliding time can be monitored during tests to determine progressive change.

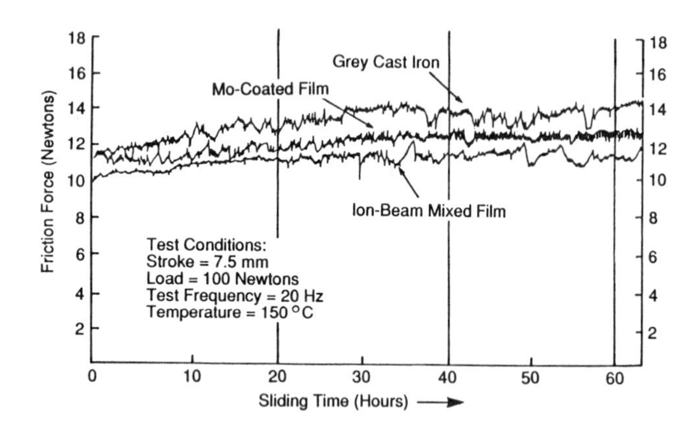

Fig. 5. A comparison of Cameron-Plint friction traces for different surface coatings under lubricated sliding conditions.

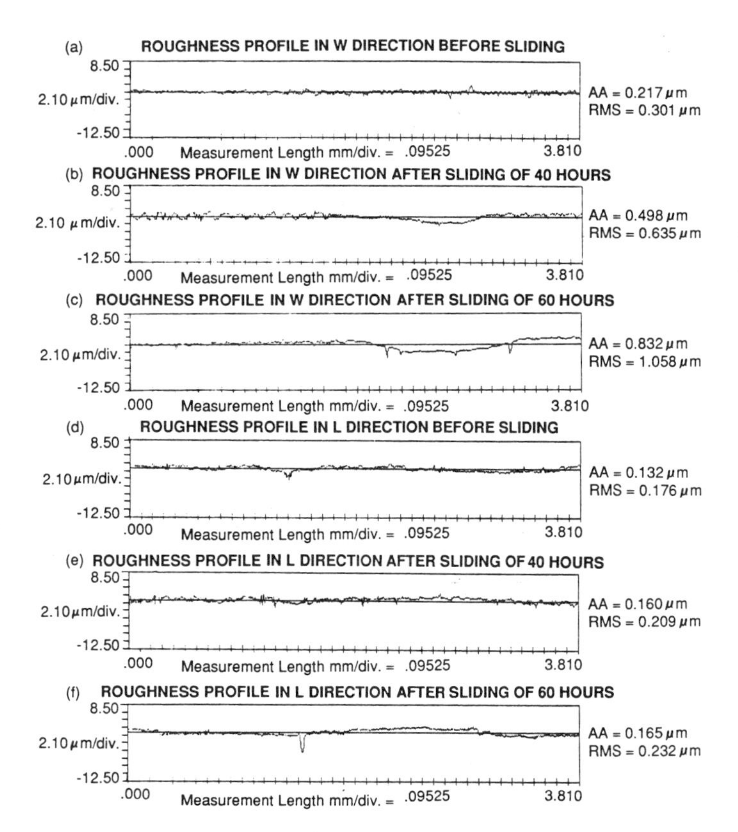

Fig. 6. Cast iron cylinder bore surface profiles and average roughness (AA and RMS) measurements for both directions. (L, sliding direction, W, across sliding) a,b,c: before, 40, and 60 hrs - across sliding d,e,f: before, 40, and 60 hrs - sliding direction

In addition, this bench test using high frequency and higher loads (severe conditions) as shown in Figure 5 can be useful to evaluate the durability of different surface coatings [1]. Both the uncoated and in-situ coated plates [1] can also be analyzed by scanning electron microscopy (SEM) and X-Ray Photoelectron Spectroscopy (XPS). As shown in Figure 7a, the plate without surface coatings was characterized by a wide wear track after sliding for sixty hours in an engine lubricant at 150°C. The wear indentations were deep and rough as described previously. Based on SEM analyses of the boundary of the wear track shown in Figure 7a, severe pitting and fatigue wear occurred during sliding. A few microcracks were also observed in the wear track. In order to analyze the worn area (the rectangular box in Figure 7a) in detail, a micrograph using higher magnification is also shown. In contrast, the plate subjected to in-situ coating, showed a film deposited on the wear track and a much smoother surface (based on surface roughness and profiles measurements). In addition, the micrograph of the wear track shown in Figure 7b indicates a protective film without any evidence of cracking or pitting wear. During the sliding process, surface coatings promoted wear resistance on the wear track.

Besides the surface analyses on these plates, this bench machine can also monitor the contact resistance [2,4]. Generally, the contact resistance during sliding fluctuates. This gives some qualitative indication of the degree of asperity interaction, since such events produce a great reduction in the electrical resistance of the contact. By using the R.M.S. of the voltage signal displayed on the oscilloscope, a continuous record of the degree of asperity contact can be made. This machine can be used to investigate the friction, lubrication, and wear behavior of materials using very simple specimen geometries and on components taken from a practical situation.

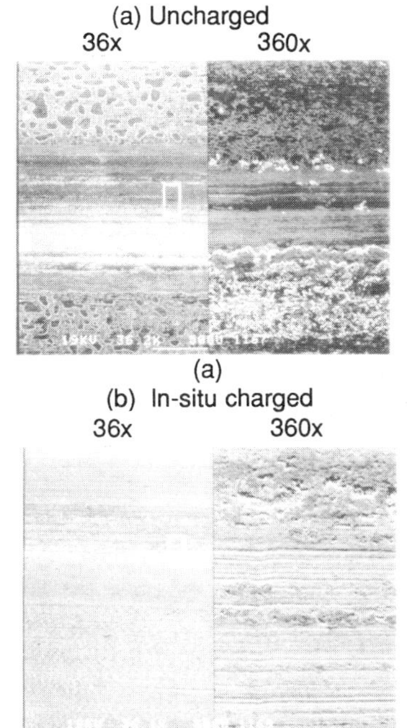

Fig. 7. A comparison of SEM micrographs taken at the boundary of the wear tracks after sliding in a lubricant for 60 hours: (a) uncoated; (b) coated plate.

EMA-LS9. The EMA-LS9 (Electro-Mechanical Associates,Inc. Model LS9) bench tester was developed in recent years in the laboratory of several of the authors, and is an outgrowth of studies on piston and ring friction in the environment of the low heat rejection diesel [6]. The essential features of this new tester are shown in Fig 8. The design philosophy was to simulate the ring and liner wear under the most severe engine condition. This is near top center combustion; characterized by high-temperature, high-load and low-speed. Here lubrication is of the boundary or mixed type, and friction is relatively high. Typically a speed less than 700 rpm and a force representative of a high maximum cylinder pressure is used. Test temperature and oil feed rates attempt to match engine conditions. Typically, a 20 hour test produces measurable wear. This tester reciprocates a clamped, full-size piston ring mounted in a disk shaped holder. The ring is mounted with the gap in the nominally closed position, thereby substantially matching the curvature of ring and liner. Segments of liner are forced to bear against the ring by the action of an air cylinder. An air atomized, metered oil spray is directed toward each of the two wear interfaces. The friction force is determined independently at each arm pivot by strain gauges and the normal force is calculated from the air cylinder pressure. The friction coefficient is calculated from the force ratio as a function of stroke, and may be monitored continuously. The wear interfaces are placed in a heated oven. A process control computer permits extended operation at a variety of pre-programmed speeds, loads, temperatures and oil feed conditions. Test parameters are monitored and key information is displayed and stored on disk as the test proceeds.

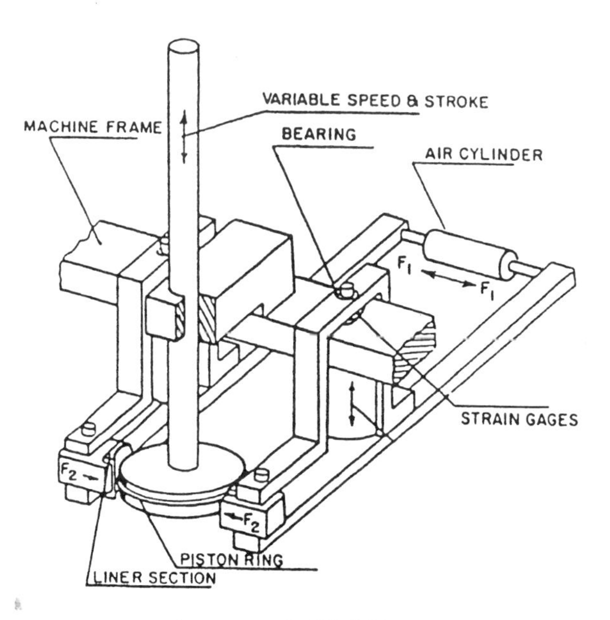

Fig. 8. Schematic of EMA-LS9 wear tester.

A typical result showing friction coefficient versus time for one lubricated ring, SPF-280 and production liner combination run under "normal" conditions for 20 hours is shown in Fig. 9. Both right and left side wear results are superimposed. The close similarity side to side in this test indicates that the test conditions and specimens were the same on each side.

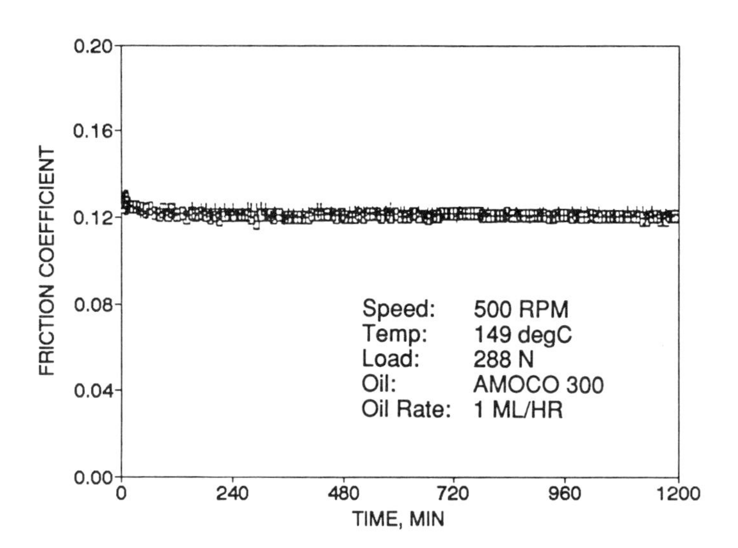

Fig. 9. Friction coefficient results for a typical 20 hour lubricated wear test for ring coating SPF-280. Right and left side results are overlaid.

Replicate wear results for three ring coatings are shown in Figure 10. Left to right the coatings are: SPF-251, a plasma-sprayed, molybdenum-based alloy coating which has been used in production for several years, X ctg A, a developing coating which was rejected because it produced excessive bore wear in engine tests, even though the coating itself was quite durable, and SPF-280, a new plasma-sprayed, chrome carbide coating developed for high wear in severe duty applications. Figure 11 shows a typical before and after test ring profile. The EMA-LS9 has been run to simulate scuffing by programming the control computer to ramp-up the load. While not shown, the liner segment wear scars may be evaluated as well as those of the rings. Typically, the liner shows more wear at the 2.5 cm stroke ends than at the mid-point. This is similar to the observed greater wear of engine liners at the dead centers.

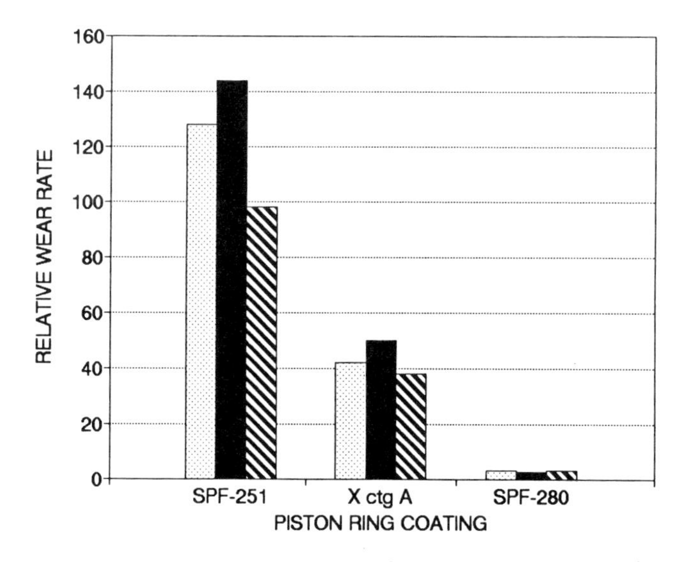

Fig. 10. Wear results based on three ring coatings, three tests each, tested under conditions of Fig. 8.

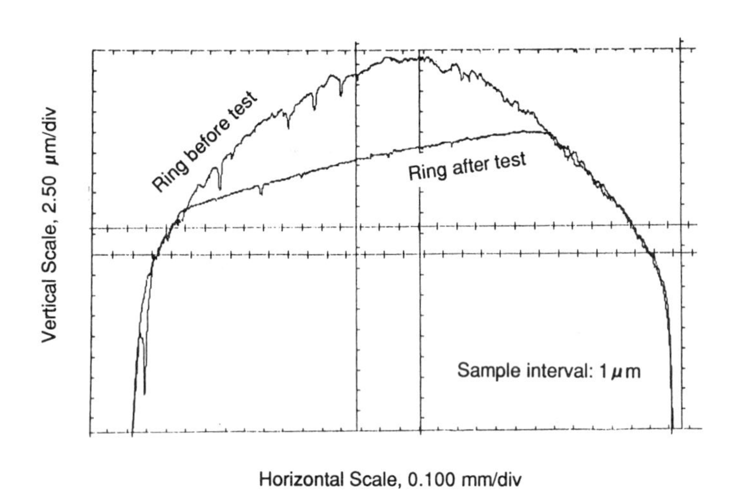

Fig. 11. Ring profile before (top profile) and after (bottom profile) test.

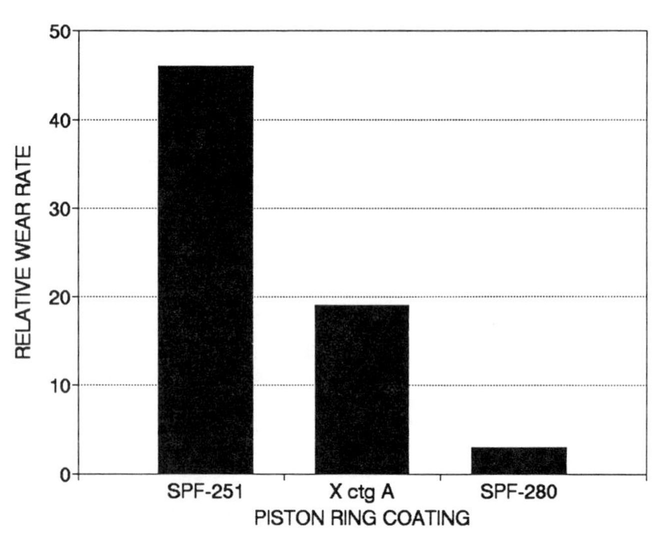

Fig. 12. Average LFW-1 wear rate results for three ring coatings.

Table 3
Correlation of Bench and Engine Wear Tests

Comparison	R	R^2	Significance Level
LFW-1 vs Gasoline Engine	0.87	76%	0.33
LFW-1 vs Diesel Engine	0.96	92%	0.04
EMA-LS9 vs Gasoline Engine	0.95	90%	0.20
EMA-LS9 vs Diesel Engine	0.99	99%	0.02

IV CORRELATION WITH ENGINE TESTS

As part of a piston ring coating development program, several sprayed coatings were evaluated on the LFW-1 and EMA-LS9 testers. The most promising coatings were then evaluated in engine dynamometer wear tests. Fig. 12 shows wear results from the LFW-1 tester for the three coatings. These are the averages of 2 - 4 tests. Results for these three materials evaluated with the EMA-LS9 were shown in Fig. 10. For correlation calculations, the average of the three tests was used. Gasoline engine wear test data were obtained also for the three coatings. In comparing the performance of the coatings between the bench testers and the engine, results were normalized with respect to the production SPF-251 coating. A wear rate factor was calculated for individual engine tests by comparing the SPF-251 wear rate in that test to the overall average wear rate for SPF-251. Wear rates of the other coatings are normalized by multiplying their wear rate by the derived SPF-251 wear factor. Statistical correlations of the LFW-1 and EMA-LS9 with the engine test results are shown in Table 3. It is seen that both the LFW-1 and the EMA-LS9 tester correlate with the engine tests. The correlation between the EMA-LS9 and the engine is particularly good.

V CONCLUDING REMARKS

Bench testing offers a rapid and low cost alternative to engine tests for screening candidate power cylinder components. With selected equipment and test procedures, the findings of this paper indicate that reasonable correlation may be expected between bench and engine tests. In the future, there will be an even greater need for improved bench testers which can even more closely correlate with engine tests. Of course there continues to be a need for more data to demonstrate the correlations.

VI REFERENCES

1. Tung, S. C., and Wang, S. S., "In-Situ Electro-Charging for Friction Reduction and Wear Resistant Film Formation", Presented in the STLE Tribology Conference (STLE Preprint No. 90-TC-6B-1), Toronto, October 7-10, 1990.

2. Plint, M.A., and Allison-Greiner, A.F.,"Routine Engine Tests - Can We Reduce Their Number?," Petroleum Review, July 1990, pp.368-370.

3. Kanakia, M.D., Cuellar Jr., J.P., and Lestz, S.J., "Development of Fuel Wear Tests Using the Cameron-Plint High Frequency Reciprocating Machine," Belvoir Fuels and Lubricants Research Facility, SwRI Report No. BFLRF 262, May 1989.

4. Plint, A.G., and Plint, M.A., "Test Procedures for Rapid Assessment of Frictional Properties of Engine Oils at Elevated Temperatures," Trib. Int., 17(4), 1984, pp. 209-213.

5. Uras, H.M., and Patterson, D.J., "Oil and Ring Effects on Piston-Ring Assembly Friction by the Instantaneous IMEP Method," SAE Paper No. 850440, 1985.

6. Slone, R., et. al., "Wear of Piston Rings and Liners by Laboratory Simulation," SAE Paper 890146, 1989.

AUTHOR INDEX